AF783143

CONTENTS

JULY 23, 2004

OPENING STATEMENT

Page

The Honorable Chris Cannon, a Representative in Congress From the State of Utah, and Chairman, Subcommittee on Commercial and Administrative Law 1

WITNESSES

Mr. Robert Pepper, Ph.D., Chief, Policy Development, Office of Strategic Planning and Policy Analysis, Federal Communications Commission
- Oral Testimony 63
- Prepared Statement 64

Mr. John Langhauser, Esq., Vice President, Law, and Chief Counsel, Consumer Services Group, AT&T Corporation
- Oral Testimony 72
- Prepared Statement 74

Mr. Stephen M. Cordi, Esq., CPA, Deputy Comptroller for the Maryland Comptroller of the Treasury, State of Maryland
- Oral Testimony 78
- Prepared Statement 80

Mr. James Kirkland, Esq., General Counsel and Senior Vice President, Covad Communications Group, Inc.
- Oral Testimony 84
- Prepared Statement 85

LETTERS, STATEMENTS, ETC., SUBMITTED FOR THE HEARING

Policy Paper of the National Cable and Telecommunications Association, "Balancing Responsibilities and Rights: A Regulatory Model for Facilities Based VoIP Competition" 3

Letter from William E. Moschella, Assistant Attorney General, U.S. Department of Justice, Office of Legislative Affairs to the Honorable Chris Cannon 51

Letter from the National Governors' Association, the Council of State Governments, the National League of Cities, the U.S. Conference of Mayors and the National Association of Counties to the Honorable F. James Sensenbrenner, Jr., and the Honorable John Conyers, Jr. 95

APPENDIX

MATERIAL SUBMITTED FOR THE HEARING RECORD

Memorandum of Understanding between the National Emergency Number Association (NENA) and the Voice on the Net Coalition 110

White Paper of the Voice on the Net Coalition, "Unleashing the Full Promise and Potential of Internet Voice Communication, Vast Benefits: lower prices, better jobs, and improved ways to communicate 111

REGULATORY ASPECTS OF VOICE OVER INTERNET PROTOCOL (VoIP)

HEARING

BEFORE THE

SUBCOMMITTEE ON
COMMERCIAL AND ADMINISTRATIVE LAW

OF THE

COMMITTEE ON THE JUDICIARY
HOUSE OF REPRESENTATIVES

ONE HUNDRED EIGHTH CONGRESS

SECOND SESSION

JULY 23, 2004

Serial No. 102

Printed for the use of the Committee on the Judiciary

Available via the World Wide Web: http://www.house.gov/judiciary

---

U.S. GOVERNMENT PRINTING OFFICE

95–009 PDF WASHINGTON : 2004

For sale by the Superintendent of Documents, U.S. Government Printing Office
Internet: bookstore.gpo.gov Phone: toll free (866) 512–1800; DC area (202) 512–1800
Fax: (202) 512–2250 Mail: Stop SSOP, Washington, DC 20402–0001

COMMITTEE ON THE JUDICIARY

F. JAMES SENSENBRENNER, Jr., Wisconsin, *Chairman*

HENRY J. HYDE, Illinois
HOWARD COBLE, North Carolina
LAMAR SMITH, Texas
ELTON GALLEGLY, California
BOB GOODLATTE, Virginia
STEVE CHABOT, Ohio
WILLIAM L. JENKINS, Tennessee
CHRIS CANNON, Utah
SPENCER BACHUS, Alabama
JOHN N. HOSTETTLER, Indiana
MARK GREEN, Wisconsin
RIC KELLER, Florida
MELISSA A. HART, Pennsylvania
JEFF FLAKE, Arizona
MIKE PENCE, Indiana
J. RANDY FORBES, Virginia
STEVE KING, Iowa
JOHN R. CARTER, Texas
TOM FEENEY, Florida
MARSHA BLACKBURN, Tennessee

JOHN CONYERS, Jr., Michigan
HOWARD L. BERMAN, California
RICK BOUCHER, Virginia
JERROLD NADLER, New York
ROBERT C. SCOTT, Virginia
MELVIN L. WATT, North Carolina
ZOE LOFGREN, California
SHEILA JACKSON LEE, Texas
MAXINE WATERS, California
MARTIN T. MEEHAN, Massachusetts
WILLIAM D. DELAHUNT, Massachusetts
ROBERT WEXLER, Florida
TAMMY BALDWIN, Wisconsin
ANTHONY D. WEINER, New York
ADAM B. SCHIFF, California
LINDA T. SÁNCHEZ, California

Philip G. Kiko, *Chief of Staff-General Counsel*
Perry H. Apelbaum, *Minority Chief Counsel*

Subcommittee on Commercial and Administrative Law

CHRIS CANNON, Utah *Chairman*

HOWARD COBLE, North Carolina
JEFF FLAKE, Arizona
JOHN R. CARTER, Texas
MARSHA BLACKBURN, Tennessee
STEVE CHABOT, Ohio
TOM FEENEY, Florida

MELVIN L. WATT, North Carolina
JERROLD NADLER, New York
TAMMY BALDWIN, Wisconsin
WILLIAM D. DELAHUNT, Massachusetts
ANTHONY D. WEINER, New York

Raymond V. Smietanka, *Chief Counsel*
Susan A. Jensen, *Counsel*
Diane K. Taylor, *Counsel*
James Daley, *Full Committee Counsel*
Stephanie Moore, *Minority Counsel*

REGULATORY ASPECTS OF VOICE OVER INTERNET PROTOCOL (VoIP)

FRIDAY, JULY 23, 2004

HOUSE OF REPRESENTATIVES,
SUBCOMMITTEE ON COMMERCIAL
AND ADMINISTRATIVE LAW,
COMMITTEE ON THE JUDICIARY,
Washington, DC.

The Subcommittee met, pursuant to call, at 10:07 a.m., in Room 2137, Rayburn House Office Building, Hon. Chris Cannon (Chair of the Subcommittee) presiding.

Mr. CANNON. The Subcommittee will please come to order.

I want to thank Mr. Chabot for being here with us this morning and helping us get this started. We are out of session, and this is an extraordinarily kind thing for him to do. We consider today the regulatory aspects of a technology that is fundamentally changing the communications industry. That is voice over Internet protocol or VoIP telephony.

As most of us know, VoIP allows the user to make telephone calls using a broadband Internet connection rather than a regular or analog hard-switched telephone line. While VoIP has been available in various forms since about 1995, the creation of new IP services and the increasing penetration of broadband into the residential markets has spurred significant growth in the industry. New and established telephony providers alike now offer various kinds of VoIP, and the service is no longer limited.

According to one estimate, the number of VoIP lines will be 4.2 million by 2007, and I suspect, personally, that that is a dramatic underestimation. At issue is whether VoIP telephony should be regulated and, if so, to what extent. VoIP represents a unique concept to regulators because it does not conform to the current regulatory paradigm which reflects the legacy system of public switched telephone network or PSTN. VoIP differs from this end-to-end telephony, because it converges services that have historically been unregulated information services and regulated telecommunications services.

The FCC's task in this regard is no minor feat. Indeed, FCC Chairman Michael Powell has stated that VoIP promises the "most important shift in the entire history of modern communications since the invention of the telephone." While the FCC first addressed the regulatory treatment of VoIP with respect to universal service in 1998, it has yet to do so in a comprehensive manner. We look forward to discussing the FCC's progress toward the establishment of a definitive framework for VoIP.

At the same time, understanding the enormous benefits of VoIP to businesses and consumers alike, prompt action is necessary that will promote rather than undermine the development of this technology. Indeed, time is of the essence for Federal guidance. Several States have launched legal or regulatory proceedings addressing VoIP, calling into question whether VoIP should be subject to State taxation or whether Federal preemption is more appropriate.

We have the opportunity today to consider those issues relevant to the development of a thoughtful yet timely approach to the regulation of VoIP from those who know the subject matter extremely well. The Subcommittee maintains jurisdiction over the Administrative Procedure Act and has a long history of providing effective oversight of the Federal administrative process by conducting hearings into regulatory practices at Federal agencies. For example, the Subcommittee has examined in hearings the FCC's regulations concerning license transfers, rules noticed by the Federal Reserve Board and Treasury Department concerning the authority to monitor banking activities and the role of Congress in monitoring administrative rulemaking. Furthermore, the Subcommittee has legislative and oversight responsibility for issues of State taxation affecting interstate commerce, which is a central issue in this debate.

I now turn to my distinguished colleague, Mr. Chabot, for any opening statement he may wish to make.

Mr. CHABOT. No.

Mr. CANNON. Thank you. The gentleman's entire statement will be placed in the record.

I ask unanimous consent that Members have five legislative days to submit written statements for inclusion in today's hearing record.

Hearing none, so ordered.

Mr. CANNON. I ask unanimous consent for the inclusion of two matters into the record. I have for inclusion in the hearing record a policy paper from the National Cable and Telecommunications Association concerning facilities-based VoIP competition and also a letter from the Department of Justice concerning the Communications Assistance for Law Enforcement Act, CALEA. This letter submits that CALEA must be considered when VoIP regulation is discussed. Without objection, these documents will be included into the record.

[The information referred to follows:]

July 22, 2004

The Honorable Chris Cannon
Chairman
Subcommittee on Commercial and Administrative Law
Committee on the Judiciary
United States House of Representatives
Washington, DC 20515

Dear Mr. Chairman:

In advance of tomorrow's hearing I wanted to share with you an NCTA White Paper we released in February concerning the regulation of Voice over Internet Protocol (VoIP) services. The paper, *Balancing Responsibilities and Rights: A Regulatory Model for Facilities-Based VoIP Competition*, describes how federal and state policymakers and regulators must affirmatively promote VoIP services as an important policy objective and adopt a predominantly deregulatory approach to VoIP services.

While much of the public policy discussion has centered on the appropriate regulatory classification of VoIP services, the paper instead focuses on the responsibilities and rights that are appropriate for facilities-based competitors offering VoIP services.

We suggest that instead of assigning a specific regulatory "box" or classification to VoIP, policymakers focus on the responsibilities and rights appropriate for a facilities-based provider of VoIP services. VoIP service providers must assume certain fundamental regulatory responsibilities including consumer protections of general applicability; assistance to law enforcement (including the principles outlined in the Communications Assistance for Law Enforcement Act); 911/E911 services and access for the disabled; contributions as appropriate to the Universal Service Fund; and participation in intercarrier compensation. VoIP providers may undertake other responsibilities on a voluntary basis but they should not be imposed.

The industry also believes that in order to provide service, VoIP providers must be accorded certain rights. These rights include interconnection to the publicly-switched network, number portability, access to universal service support and intercarrier compensation, and access to rights-of-way and other facilities without incremental fees. The regulatory classification under which this set of responsibilities and rights is established is important, though ultimately less important than those responsibilities and rights being established in a minimally regulatory framework.

We hope you find this paper useful, and look forward to working with you to make this framework a reality.

Sincerely,

Enclosure

Balancing Responsibilities and Rights:
A Regulatory Model for Facilities-Based VoIP Competition

An NCTA Policy Paper

February 2004

NCTA
NATIONAL CABLE & TELECOMMUNICATIONS ASSOCIATION

Balancing Responsibilities and Rights: A Regulatory Model for Facilities-Based VoIP Competition

An NCTA Policy Paper

NCTA
NATIONAL CABLE & TELECOMMUNICATIONS ASSOCIATION

Table of Contents

Introduction and Executive Summary...i

I. What is VoIP? 1

II. The Opportunity Presented by Facilities-Based VoIP Services ..5

III. The Regulatory Challenge of Deploying New Services.............8

IV. VoIP Regulatory Proceedings in the States............................15

V. NCTA's Approach: Balancing Responsibilities and Rights.......20

VI. The Responsibilities and Rights of VoIP providers..................21

Public Health and Safety...22

Universal Service...22

Intercarrier Compensation...24

Consumer Protection........... 25

Inappropriate Legacy Utility Requirements......... 25

Rights of VoIP Providers...26

VII. Regulatory Restraint and Regulatory Classification................28

Conclusion...39

A Regulatory Model for Facilities-Based VoIP Competition

Introduction and Executive Summary

Today, most American households do not have a choice of facilities-based local telephone service providers. They have not realized the benefits of such choices despite nearly a decade of efforts by lawmakers and regulators to promote facilities-based competition in the local telephone marketplace. Although some cable companies are providing an alternative with circuit switched telephone service, with the deployment of cable-based Internet Protocol ("IP") phone services, customers will enjoy new options for a full suite of facilities-based voice services.

Forms of non-facilities-based Voice over Internet Protocol ("VoIP") service exist today, but they generally do not offer the reliability and quality that consumers have come to expect from "plain old telephone service" ("POTS") offered by incumbent local exchange companies ("ILECs") and most competitive local exchange companies ("CLECs"). Cable communications companies are working to introduce a new generation of phone services that will offer the flexibility and economy of IP technology (*i.e.,* the shared transmission of voice, data, and video information via a managed network) and the reliability and quality of service that consumers desire. Importantly, VoIP services delivered over a broadband cable network will, over time, provide wide-scale residential phone competition that is both *facilities-based* and *sustainable*.

The cable industry is excited about the consumer benefits and business opportunities that VoIP services will create, and the industry is devoting capital, personnel, and other resources to make facilities-based VoIP services a marketplace reality. Resources and the state of technological development, however, are not the only factors that will affect the availability of VoIP services. Regulatory uncertainty – and the potential for application of unnecessary or overly burdensome regulation – will also affect whether, when, and how VoIP services are deployed.

The Internet and information services generally have succeeded, in large measure because of regulators' prescient and courageous decision, made more than two decades ago, to promote competition in interstate information services and to fence them off from unnecessary federal and state regulation. Commercial mobile radio services ("CMRS") have similarly been the subject of pro-competitive and deregulatory policies, again with salutary results in terms of investment, speed of innovation, and competition. Unfortunately, this has not generally been the case for CLECs. Although some states have adopted a hands-off approach to regulating new entrants, many states have imposed varying levels of traditional telephone regulation on those new entrants. It is unknown how the costs of this regulation have affected the willingness of companies to commit risk capital and provide competitive alternatives. Establishing a clear legal framework that promotes the emergence of VoIP services and ensures their freedom from unnecessary regulation can have equally beneficial results for the development of telephone competition, particularly in the residential mass market.

Much of the public policy discussion surrounding VoIP has centered on the appropriate regulatory classification of such services. Such an approach, however, has several shortcomings, as each regulatory category carries with it a history of regulatory assumptions that may or may not be appropriate for new technologies such as VoIP and the services they spawn. For that reason, this policy paper chooses instead to describe the cable industry's vision for a regulatory approach that will lead to efficient and rapid deployment of facilities-based VoIP services. We describe the public policy objectives that should be pursued to encourage the growth of VoIP services. We propose a regulatory roadmap that: (1) assigns to VoIP service providers vital *responsibilities*; (2) discusses certain responsibilities that VoIP service providers may undertake on a voluntary basis, but which should *not* be imposed upon them; and (3) identifies *rights* that are essential for VoIP service deployment. We also establish a baseline definition as to *which* VoIP services should have such rights and responsibilities. In doing so, we suggest that such an approach be applicable to new entrant VoIP service providers based upon the precise nature of the services they provide, regardless of whether they provide those services over their own facilities or the facilities constructed by others.

Protecting VoIP services from unnecessary regulation does not require that important public policies be neglected. Even under a generally deregulatory regime, any VoIP service that meets a baseline test as proposed herein[1] can, and should, meet certain public policy responsibilities and requirements such as the principles set forth in the Communications Assistance for Law Enforcement Act ("CALEA"), the offering of 911/E911, access for the disabled, and appropriate contributions to universal service. But the overall direction of public policy should be toward a deregulatory environment in which even the most vital public policy objectives are secured through the lightest possible regulation, so as not to forestall the many benefits of these new services.

Similarly, there are a number of legacy utility requirements that should not be imposed on VoIP service providers. Most such requirements date from the era of a single provider of phone service and are inappropriate for competitors using nascent technologies that offer alternatives to incumbent providers. In particular, a number of legacy requirements relate to billing, payment, credit and collection, and quality of service standards. Competitive marketplace forces, rather than prescriptive rules, can address these issues much more effectively for non-incumbent providers of VoIP services. Regulators should make a comprehensive effort to review and eliminate such regulatory requirements for VoIP services.

VoIP service providers, particularly facilities-based providers, do, however, require certain rights irrespective of whether the provider's service is ultimately determined to be an "information service," a "telecommunications service," or another type of service. These rights relate generally to interconnection and the exchange of traffic, the right to obtain telephone numbers and have them published in telephone directories, the right to access the facilities and resources necessary to provide VoIP customers with full and efficient 911/E911 services, the right to be compensated fairly for terminating traffic delivered from other entities and the right to non-discriminatory

---

[1] The proposed four-prong test requires that a VoIP service (1) use North American Numbering Plan ("NANP") resources, (2) receive calls from – or terminate them to – the public switched telephone network ("PSTN"), (3) represent a possible replacement for POTS, and (4) use Internet Protocol transmission between the service provider and the end user customer, including use of an IP terminal adapter and/or IP-based telephone set.

access to universal service support. In addition, facilities-based VoIP providers need access to poles, ducts, conduits and rights-of-way, regardless of the ultimate regulatory classification of VoIP services.

In the final analysis, facilities-based VoIP services can be the breakthrough that fulfills the vision of the Telecommunications Act of 1996[2] ("1996 Act") for vast numbers of residential consumers. The cable industry stands ready to play a lead role, just as it has done in making residential broadband Internet service a widespread and desirable service. This breakthrough will occur most rapidly and ubiquitously if federal and state policymakers and regulators affirmatively promote VoIP services as an important policy objective and adopt a predominantly deregulatory approach to VoIP services.

[2] Telecommunications Act of 1996, Pub.L. No. 104-104, 110 Stat. 56 (1996).

I. What is VoIP?

VoIP is the convergence of voice and data into a single bitstream, which enables the provision of innovative offerings that integrate the two in ways not possible using traditional circuit-switched technology. Voice communications are digitized into data packets and routed in that form over either managed IP networks and/or over the public Internet to the desired location using IP addressing. As such, VoIP, in and of itself, is not a service. Rather, VoIP is a technology that allows voice traffic to be packetized and transported or routed over privately managed networks as data packets. Because the vast majority of telephone subscribers continue to be served by incumbent LECs on the public switched telephone network ("PSTN"), most VoIP-based calls made today continue to traverse, at some point, the PSTN. As VoIP-based services become more prevalent, however, the technology will eliminate the need for both traditional circuit switching and the public switched telephone network ("PSTN").

In traditional circuit-switched telephony networks, a dedicated path, or channel, is opened between the parties participating in the call. No other traffic can pass over that channel while the call takes place. This dedicated channel remains open until the parties terminate the call, thus freeing up the channel for use in another call. In VoIP telephony – as with other IP-based services – dedicated circuits are not used. Multiple conversations are sent over the same channel as separate streams of data packets. When there is a lull in any particular conversation, other data packets can be carried over the same portion of the network, thus making the network more efficient than a traditional circuit-switched network. In technical terms, VoIP uses the network more efficiently because it combines, or multiplexes, multiple sets of data over the same physical path.[3]

VoIP is an attractive technological approach for cable system operators who have already entered the local telephone market as well as those offering voice

---

[3] *See VoIP – the Enabler of Real Telecom Competition*, Goldman Sachs Global Equity Research Jul. 7, 2003 at 3.

services for the first time. Compared to circuit-switched telephony, VoIP may result in lower (though still significant) rollout costs, increased flexibility, and more innovative and advanced services. More specifically, VoIP allows a provider to avoid the huge capital expenditures and investments needed to purchase and install circuit switches. Furthermore, VoIP utilizes data paths that the cable industry has already invested in and built. These existing paths facilitate easy software changes and additions to service packages, as well as innovative combinations of voice, data, and fax services.

As with many other technical pursuits, standardization is important to VoIP. Cable companies want to be able to purchase equipment from various vendors, and to know that the equipment will be interoperable. To that end, CableLabs, the industry's research consortium, has been involved in developing uniform technical specifications for many years, including a successful effort to develop cable modem technical specifications. The Data Over Cable System Interface Specification ("DOCSIS") is also the underlying specification for a CableLabs project known as PacketCable. Very simply, PacketCable is a common platform and set of interoperable interface specifications for delivering advanced, real-time multimedia services, including not only VoIP, but also multimedia conferencing, interactive gaming, and other multimedia applications. The VoIP specifications are written to do exactly what today's analog, circuit-switched phone network does, from dial tone to ring tone. But unlike other VoIP specification efforts that address only individual portions of how to make an IP phone call, PacketCable addresses the entire journey.

The term "VoIP" encompasses these, as well as many other services, ranging from voice-enabled instant messaging and chat and voice-enabled gaming (such as Xbox Live) to services which replicate POTS. In many instances, "VoIP" will simply support a voice application or software application. Among the services that some cable operators are considering are "unified" messaging (whereby users have a single message platform for e-mail, voicemail, faxes, and the like); personal portals; caller ID on television sets; talking email; and customized dial-tones and greetings. VOIP may also make possible advanced video conferencing services including a combination of voice, video, and data delivery. Furthermore, with VoIP, some consumers may eventually be able to use the Internet from any location and instruct a home phone to forward calls to another phone number or listen to voicemail via the Internet from any location. Or, in an example offered by FCC Chairman Michael Powell, because "[VoIP] can be readily integrated with other computing systems ... you make an Internet call to a doctor's office to make an appointment. The doctor's system calls up your medical records, your medications, and your last visit and instantly displays them. It also brings up the appointment times available, allows you to select one and then calls you back, or sends a text message to your cell phone, the day before the appointment to remind you."[4]

Even among those VoIP services that are "phone-like" there are significant differences. For example, the IP data packets used by services from some of the currently well-known providers, such as Vonage, travel over the public Internet. Facilities-based cable offerings, in contrast, will be able to transport IP data packets over their private managed IP networks with end-to-end quality of service monitoring (while still interconnecting with the PSTN as necessary). Moreover, with a cable-based VoIP service, it is possible to offer a robust VoIP service to a customer that does not subscribe to high-speed Internet access service. At least one cable company is currently offering its VoIP product to customers who do not subscribe to high-speed Internet access.

The VoIP services of particular concern in this paper might be more properly referred to as "IP Phone" services – those that in some ways mimic traditional telephone service. It appears, however, that the term "VoIP" has come to commonly

---

[4] See *The Age of Personal Communications: "Power to the People"*, Remarks of FCC Chairman Michael K. Powell Before the National Press Club, Washington D.C. (Jan. 14, 2004), available at http://www.fcc.gov/commissioners/powell/spmkp011404.pdf. In a further example "[s]imilar potential rests with police and fire response systems. The 911 system is vital in our country, but it is limited functionally. In most systems, it primarily identifies the location from which the call was made. But an Internet voice system can do more. It can make it easier to pinpoint the specific location of the caller in a large building. It might also hail your doctor, and send a text or Instant Message alert to your spouse."

refer to these phone-like services and thus this paper will use that term. It is important to recognize, however, that there remain distinctions among the type of VoIP-based services discussed herein. Indeed, nomenclature may be part of the very debate over VoIP policies. As discussed in more detail below, however, the cable industry believes that regulatory distinctions should be drawn based upon the type of services being provided by new entrant VoIP providers and not whether, for example, the service provider routes calls over the "Internet" or owns the facilities over which it routes calls. Few would argue, for example, that applications, or devices, where voice functionality is ancillary to the actual purpose of the service or device and where such applications do not fall within the specific VoIP service defined herein—as in voice-enabled gaming—should be regulated in the same manner as a traditional phone service.

Given these many distinctions, policymakers should establish a baseline test to determine whether an IP-based voice service should be subject to any regulation at all[5] (as described in Section VI). Specifically, that test should be based on whether the VoIP service in question has the following characteristics:

1. it makes use of North American Numbering Plan ("NANP") resources;
2. it is capable of receiving calls from or terminating calls to the public switched telephone network ("PSTN") at one or both ends of the call;
3. it represents a possible replacement for POTS; and,
4. it uses Internet Protocol transmission between the service provider and the end user customer, including use of an IP terminal adapter and/or IP-based telephone set.[6]

---

[5] While it may, however, be warranted to require applications that do not meet this baseline test to provide assistance to law enforcement for security reasons, there appears to be no justification for imposing traditional telephone regulation upon such applications.

[6] See *Federal-State Joint Board on Universal Service*, Report to Congress, 13 FCC Rcd. 11501 (1998) ("*Stevens Report*"). In particular, the report established a four-part test, with the fourth prong relating to equipment. Given the advances in customer premises equipment, and the blurring of the lines

IP applications such as voice communications overlaid on video gaming or video chat, which do not have the characteristics of the first three prongs above, should not be subject to regulation, much less traditional telecommunications regulation. Such applications generally would not use NANP resources nor would they have the ability to receive calls from or terminate them to the PSTN. The services covered by the four-prong test, as with others that are facilities-based, would fulfill the promise of the 1996 Act in promoting the goal of greater residential competition. Services lacking characteristics of the fourth prong (i.e., lacking an IP based connection to the end user), are not addressed by this VoIP proposal.

II. The Opportunity Presented by Facilities-Based VoIP Services

Over the years, and particularly since the 1996 Act, a consensus has evolved that American consumers will reap the greatest benefits from communications policies that encourage industry investment, foster technological innovation and service deployment, and increase consumer choices. To that end, Congress, in the 1996 Act, declared its intention to promote competition and to eliminate unnecessary regulation.[7] These goals – investment, innovation, choice, competition, and deregulation – should be the primary reference points for policymakers' response to emerging VoIP services.

A central objective of the 1996 Act was to introduce facilities-based competition into the local phone services market.[8] Nearly eight years later, competition in the local

---

between computers and phones nearly six years later, the fourth prong in that 1998 report no longer seems germane.

7 *See 1996 Act* at preamble (stating that the purpose of the 1996 Act is to "*promote competition and reduce regulation* in order to secure lower prices and higher quality services for American telecommunications consumers and encourage the rapid deployment of new telecommunications technologies") (emphasis added).

8 The FCC has explicitly found that "facilities-based competition serves the Act's overall goals." *Review of the Section 251 Unbundling Obligations of Incumbent Local Exchange Carriers*, CC Docket No. 01-338, Report and Order and Order on Remand and Further Notice of Proposed Rulemaking, FCC 03-36, at ¶ 70 (rel. Aug. 21, 2003). Specifically, "[f]acilities-based competition better serves the goal of *deregulation* because it permits new entrants to rely less on incumbent LECs' facilities and on

phone services market remains a hope rather than a reality for the vast majority of residential consumers. Although some markets enjoy the benefits of facilities-based competition from companies who have taken the risk and made the investment, this is atypical. In a majority of markets, residential consumers have no meaningful choice of facilities-based local phone service providers.

This is despite the fact that the cable industry has recognized the importance to its customers of developing robust, competitive local phone services. Companies such as Cablevision Systems Corporation, Charter Communications, Comcast Corporation, Cox Communications, Inc., GCI Cable, Inc., and Insight Communications collectively serve over 2.5 million subscribers with circuit-switched telephone service.[9] And even as these companies maintain and improve existing circuit-switched local telephone operations in their service areas, they are preparing to expand the range of service options – and the places in which those options are available – using facilities-based VoIP technologies.

In other areas where a choice exists, it typically consists of mere resale of the incumbent's services or the use of the incumbent's unbundled network elements in a combination known as "the unbundled network element platform" or "UNE-P." The

---

regulated terms for access and price. And it serves the goal of *innovation* because new facilities are more likely to have additional capabilities to provide new services to consumers and competitors' deployment of new facilities is likely to encourage incumbents to invest in their own networks. Facilities-based competition also increases the likelihood that new entrants will find and implement *more efficient technologies*, thus benefiting consumers. . . . Finally, facilities-based competition creates network redundancy, which increases *reliability* and enhances *national security*." *Id.* at n. 233 (emphasis added; internal citations omitted).

9 In the former AT&T Broadband territories, Comcast continues to offer circuit-switched telephone services in each of the 18 markets where competitive telephone service was previously offered by AT&T Broadband, and to solicit and process orders from new customers. As of the third quarter of 2003, Comcast had over 1.3 million residential phone customers (including a small number of customers from preexisting Comcast operations in Maryland, Michigan, and Northern Virginia), making it the largest residential facilities-based CLEC in the U.S. Comcast currently offers a facilities-based circuit-switched competitive choice to nearly nine million households.

Cox, a pioneer in circuit-switched cable telephony offers competitive circuit-switched telephone services to over 4 million households in 11 major markets across the country. As of the third quarter of 2003, Cox had nearly 1 million residential phone customers.

regulatory regimes of resale and UNE-P were intended, pending the emergence of facilities-based competition, primarily as transitional mechanisms. Unfortunately, the telecom industry has been mired in nearly eight years of rulemakings and litigation over the UNE regime and related provisions of the 1996 Act. What has languished, especially in the residential marketplace, is the development of the robust *facilities-based* competition that Congress believed could best provide enduring consumer benefits.

Now, however, VoIP technology offers the key to this long-awaited competition. The potential exists – by harnessing the same IP technology that is the foundation of the Internet – for a platform other than the incumbents' local exchange network to deliver telephone service on a wide scale, providing residential consumers with real choice in facilities-based local phone service. IP technology offers the additional consumer benefit of enabling third parties to utilize this new platform to provide VoIP service in competition with one another as well as with the incumbent telephone companies.

As a result of more than $84 billion of private investment in upgrades and enhancements to cable technology since 1996, cable operators are preparing to provide innovative facilities-based VoIP services in many areas – services that support 911/E911 and the principles of CALEA and are delivered via a managed network with a quality-of-service standard. VoIP regulatory policy must ensure that cable operators who invest in the platform that makes this competition possible are not disadvantaged by regulation in favor of those who use that platform to compete with cable's VoIP services. With the right regulatory framework, VoIP technology will increase industry investment, foster innovation, and provide consumers with attractive alternatives to POTS and to other communications services.

III. The Regulatory Challenge of Deploying New Services

Potential providers of any new services face the uncertainty of regulation at the federal, state and/or local level. Until now, consumers and providers have benefited from the decision by policymakers not to legislate or regulate in a manner that discourages innovation and investment in VoIP services.[10] This is particularly so at the federal level. For several years, limited forms of VoIP service have been offered without regulation. While the earliest forms of non-facilities-based VoIP service did not provide traditional phone service quality or reliability, consumers used those services to replace calls to countries with high international toll rates – with the strong encouragement of the Federal Communications Commission ("FCC").[11] Today, providers such as Vonage, ePHONE, ICG Communications, Inc., and pulver.com are providing forms of VoIP services with little or no governmental regulation.[12]

While the federal government to date has suggested it will take a "hands-off" approach to regulating VoIP, a major concern for would-be VoIP service providers is that one or more states could subject their services to existing state-specific regulatory schemes and/or establish new and equally burdensome regulations for VoIP services. State regulators have recognized the danger inherent in such an approach, as well. For

---

[10] *See, e.g., Stevens Report* (noting the FCC's desire for the VoIP industry to develop from a nascent service prior to making regulatory decisions that could stifle development: "[W]e recognize the need, when dealing with emerging services and technologies in environments as dynamic as today's Internet and telecommunications markets, to have as complete information and input as possible").

[11] *See, e.g., Rules and Policies on Foreign Participants in the U.S. Telecommunications Market*, Report and Order and Order on Reconsideration, 12 FCC Rcd. 23891 at ¶ 16 (1997) (noting that new technologies such as "Internet telephony are already putting significant pressure on international settlement rates and domestic collection rates"); *see also* Kevin Tanzillo, *FCC to Teach Old Tricks to New Dogs*, Communications News, Jul. 1, 1996 (quoting former FCC Chairman Reed Hundt: "'I think that Internet telephony will initially have the biggest impact on the price of international long-distance calls. . . . When China is more accessible to the Internet, it will come to pass that the current $4.35 per minute charge for a long-distance call to China will dissolve like spit in the wind").

[12] *See Petition for Declaratory Ruling That AT&T's Phone-to-Phone IP Telephony Services Are Exempt from Access Charges,* FCC WC Docket No. 02-361, Joint Comments of Association for Communications Enterprises, Big Planet, Inc., ePHONE Telecom, ICG Telecommunications, Inc., and Vonage Holdings Corp. (filed Dec. 18, 2002). *But see infra* Section IV (describing the efforts of some states to regulate VoIP service).

instance, Colorado PUC Chairman Gregory E. Sopkin has warned that the "nascent VoIP industry should not be subject to death-by-regulation, which could well occur by having 51 state commissions imposing idiosyncratic, inconsistent, and costly obligations." [13] (State regulatory activity is described in the next section).

The application of traditional state telephone regulations risks encumbering VoIP services with a web of costly and potentially inconsistent rules that will inevitably deter potential market entrants from offering the services, especially since the efficient multi-state rollouts of VoIP will depend on new centralized ordering, provisioning, and billing systems. Encumbrances are also possible at the local level, where at least some communities argue that *all* services delivered over cable plant should be subject to separate and duplicative municipal fees, requirements for additional permits, quality standards, privacy rules, and the like.[14] This local layer of regulation makes no sense when the new services can be offered simply by changing the pattern of signaling sent over an existing physical transmission facility, without imposing any additional burden on rights-of-way. This is precisely the situation with cable-delivered VoIP services.[15]

---

[13] Colorado's VoIP proceeding (Dkt. 03M-220T), begun in May 2003, ended based on the "legal uncertainty of whether a state may regulate VoIP services," concluding that "the most prudent course is to take no action with respect to VoIP pending FCC action." *See* TR State Newswire, *PUC ends VoIP Investigation, Sopkin voices views on VoIP,* Jan 6, 2004. "Sopkin added that VoIP shouldn't be regulated like traditional phone service. 'We should treat VoIP not as a problem, but a new opportunity for regulators to look at changing how the use of wireline infrastructure is compensated – through subsidies, intercarrier charges, and regulated rates.' The chairman called on VoIP providers to seek free market solutions to intercarrier compensation and 911 service issues, urging them to negotiate service agreements 'to show they are good corporate citizens and to show that traditional regulation is not necessary.'"

[14] *See Inquiry Concerning High Speed Access to the Internet over Cable and Other Facilities, Appropriate Regulatory Treatment for Broadband Access to the Internet over Cable Facilities,* FCC GN Dkt, Nos. 00-185, 02-52, Comments of Alliance of Local Organizations Against Preemption (filed Jun. 17, 2002).

[15] Likewise, regulators must not subject VoIP services to financial penalties in the form of high pole attachment fees. VoIP services will normally be carried over pre-existing facilities already attached to utility poles. There will be few if any new poles placed or new trenches dug, and there will be few if any new wires attached to existing poles. VoIP services delivered by cable operators will be offered by simply changing the pattern of electrical and optical signals carried over existing physical facilities already in use for other purposes (*e.g.,* delivery of video entertainment and/or high-speed connectivity to the Internet). Regulators, in considering the issue of pole attachment rates, must therefore avoid

Moreover, local micro-regulation of new services such as VoIP would stifle them. Cable operators today can be subject to dozens or even hundreds of local franchising authorities for their cable systems in a single state. Offering VoIP services would be immensely more difficult with dozens or hundreds of inconsistent regulations.

Congress, the FCC, state legislatures and commissions, and local governments all need to adopt an approach that will encourage the deployment of VoIP services in general, and of facilities-based services (VoIP and otherwise) in particular. Factors warranting emphasis in the analysis include the nascent nature of the services, the desirability of fostering, on a broad scale, a facilities-based alternative to incumbent local phone services, delays in deployment that could result from a tangle of incongruous state and local regulations, the importance of providing regulatory certainty in the near term, and the likelihood that the VoIP services of various providers will include differing capabilities. For all these reasons, it is critical that policymakers and regulators ensure that regulation does not become an impediment to VoIP service testing, investment, innovation, and deployment.

Ultimately, however, much of the responsibility lies with the FCC. The FCC has the ability to bring states and providers together (for example, through its announced intention to issue a Notice of Proposed Rulemaking or "NPRM" on VoIP services soon) to determine on a uniform national basis which regulatory requirements are truly needed and which regulatory requirements will pose unnecessary barriers to entry and growth, as well as to articulate and enforce a suitably deregulatory (but not entirely deregulated) policy framework that allows for maximum flexibility, innovation, investment, and competition. The FCC's announced NPRM appears to have already

---

applying regulatory categories or regulatory solutions to those new and innovative services developed with other technologies in mind. Clearly, it would make no economic or policy sense for regulators to take a regulatory approach to VoIP services which would result in an unearned windfall to those who control poles merely based on a change in the pattern of optical and electrical signals carried over existing facilities and infrastructure. A change in these signals has no economic or physical impact on poles, conduits, or rights-of-way, yet it is all that is needed to offer VoIP service.

had the effect of convincing states such as California to step back from efforts to possibly regulate VoIP providers as traditional telecommunications carriers.[16]

The FCC and state regulators, in developing a policy framework, should avoid perpetuating approaches that penalize industries such as the cable industry that have been willing to assume the added financial and other risks of building and continually upgrading the physical infrastructure needed to enable delivery of VoIP services. The FCC and state regulators should instead embrace regulatory approaches that encourage deployment of that competitive infrastructure.

Notwithstanding the regulatory challenge of deploying new services, cable operators have been among the early leaders in developing facilities-based VoIP technology to serve the residential market. Current company rollouts include:

- **Armstrong** has partnered with VoIP service provider Vonage to offer Zoom phone service to cable customers throughout Armstrong's 11 cable systems, located in Kentucky, Maryland, Ohio, Pennsylvania, and West Virginia. The service is essentially a private label rebranding of Vonage service. Armstrong's residential packages range from a $24.99 product with unlimited local and regional calling and 500 minutes of long distance across the US and Canada to a $34.99 product with unlimited local and long distance calling across the US and Canada. Just as with the Vonage product, a potential Zoom customer must subscribe to broadband service and use a digital phone adapter which plugs into the DSL or cable modem (in this case a cable modem). The adapter has "[b]uilt in Quality of Service (QOS) technology [which] prioritizes your voice data over other [I]nternet traffic ..."[17]

---

[16] *See* Ben Charny, *California eases up on Net phone rules*, CNET News.com (Jan. 5, 2004), *available at* http://news.com.com/2100-7352-5135188.html?tag=guts_lh_7352.

[17] *See* http://www.zoom-phone.com/features.php or http://www.vonage.com/features.php.

- **Cablevision** launched Optimum Voice, a digital voice-over-cable service, in the fourth quarter of 2003 throughout its New York City metropolitan service area of more than 4 million homes (which includes Bronx, part of Brooklyn, Long Island and the Lower Hudson Valley as well as southern Connecticut and northern New Jersey). Optimum Voice is currently the largest facilities-based VoIP deployment in the United States. The service provides unlimited local, regional, and long distance calling across the US (including Alaska and Hawaii) and Canada for a flat rate of $34.95 per month. It includes five customer calling features (call waiting, caller ID, call return, three-way calling and call forwarding) and E911. Currently, Cablevision is offering Optimum Voice to its more than 1 million high-speed Internet service customers. Area code and phone number assignments are based on the location of the customer's residence.
- **Charter** launched commercial VoIP service in September, 2002 in Wausau, Wisconsin and is now gearing up its marketing efforts. In addition to expanding VoIP in its Wisconsin footprint, Charter will launch VoIP service in several other markets this year.
- **Comcast**, the largest cable company with 1.3 million telephony subscribers nationwide, is currently testing VoIP near Philadelphia, Pennsylvania and plans to trial the service in several markets including Indianapolis, Indiana, and Springfield, Massachusetts in 2004. Comcast has indicated its intention to "differentiate itself from telcos with inexpensive deals on four lines, since they don't cost the provider more than one, and video enhancement of service comparable with instant messaging, Internet chat or voice mail."[18]
- **Cox** launched its first VoIP service, Cox Digital Telephone, in December 2003 in Roanoke, Virginia, representing the twelfth market in which Cox has introduced

---

[18] *See Cable VoIP Will Provide the Facilities-Based Phone,* Communications Daily (Dec. 15, 2003), at 6, quoting Comcast CEO Brian Roberts speaking at the Commonwealth Club (San Francisco).

phone service. (In its other eleven telephone markets, Cox relies on traditional circuit-switched technology.) Cox Digital Telephone subscriptions grew on the order of forty percent in 2003. In the past several years, Cox has pioneered cable telephony via circuit switched technology, gaining experience central to its VoIP launch while earning highest honors in J.D. Power and Associates' 2003 Residential Local Telephone Customer Satisfaction Study in the Western Region. Cox's telephony launch using VoIP-based technology provides customers with the same lifeline service as traditional telephone service, including E911 access and popular calling features such as call waiting, caller ID and voicemail. Cox's self-managed VoIP architecture also supports local number portability, enabling customers to switch their existing phone numbers to Cox Digital Telephone service.

According to CNET News "[s]maller markets such as Roanoke represent 19 of the 21 other markets into which Cox wants to expand its voice service. VoIP is an ideal candidate – these areas might not generate the profits necessary to validate the outlay involved with a more traditional system, Cox spokesman Bobby Amirshahi says. 'In smaller markets, it becomes a major question of whether you can justify the cost of circuit switched,' according to Amirshahi."[19]

- **GCI** has begun deployment of a hybrid VoIP/circuit switched service in Anchorage, Alaska, where it currently serves over 40 percent of the market, primarily via UNE-loop. The service being deployed is based on PacketCable standards from the customer premises to a media gateway and then uses GCI's circuit-switched facilities. As GCI transitions customers to its own loop facilities, it will be able to reduce its use of the incumbent local exchange carrier's facilities

---

[19] *See* Ben Charny, *Cox Communications Dives into VoIP*, CNET News.com (Dec. 15, 2003), *available at* http://news.com.com/2100-7352-5124440.html?tag=guts_lh_7352.

- **Time Warner Cable** launched Digital Phone, its VoIP service, to subscribers in Portland, Maine in May of 2003. By year-end 2003, Time Warner Cable had signed up more than 9,000 subscribers who pay $39.95 (for digital cable television and/or high-speed Internet subscribers) or $49.95 (for customers that do not subscribe to digital cable television or high-speed Internet services) for unlimited local and domestic long distance calling. The service includes call waiting, caller ID and call waiting ID, access to E911, and the option of local number portability. Subscribing to digital cable television or cable Internet service is not a prerequisite to purchase Digital Phone, although a potential Digital Phone subscriber must, at a minimum, subscribe to either cable television service or high-speed Internet service.

  Time Warner Cable recently launched its Digital Phone service to select customers in North Carolina and plans to offer the service by the end of 2004 in most, if not all, major markets in the 27 states it serves. This means the company's Digital Phone product should be available to nearly its entire footprint of over 11 million subscribers and over 18 million homes passed.

  In December, 2003 Time Warner Cable announced a partnership with long distance companies MCI and Sprint to assist in provisioning Digital Phone service and to use their networks to carry calls from its cable network to receiving callers served by traditional PSTN-based providers. In addition to providing long distance services, MCI and Sprint will support E911access and local number portability, permitting Time Warner Cable to continue its aggressive rollout in 2004.

As these services are deployed, cable companies continue to test and develop back-office support systems, provisioning and operational processes (including billing), and marketing programs. These efforts, and the various announced deployments, attest to the industry's belief that VoIP technology will ultimately permit cable operators to provide innovative, high-value residential local phone services at competitive prices.

Clearly, the industry is excited about and committed to the potential benefits that can result from the widespread availability of VoIP services. Yet, a broad roll-out of these services is not assured. A key factor that will affect the ability of cable companies to offer commercially viable VoIP services is the (de)regulatory framework that applies to these services, particularly the services offered in competition with incumbent providers. Where incumbent utilities offer VoIP services in their legacy franchise or service areas as substitutes for POTS services, it is important for regulators to consider whether to maintain appropriate regulatory safeguards, particularly in light of the goal of promoting facilities-based competition in the 1996 Act.

IV. VoIP Regulatory Proceedings in the States

Some states, such as Colorado, Florida, and Pennsylvania have appropriately taken a deregulatory approach to VoIP services. As described below, other states are applying existing intrastate access charge regimes to VoIP services without awaiting the outcome of FCC proceedings addressing interstate access charges. Still others have required (or are considering requiring) VoIP service providers to comply with most or all state laws and regulations that apply to traditional telephone service. Below is a brief description of the major VoIP proceedings underway in the states:

Alabama – In July 2003 a group of local exchange carriers filed a Petition for Declaratory Ruling at the Alabama Public Service Commission (the "Alabama PSC") seeking to classify VoIP providers as "transportation companies" under Alabama law, and declaring that they are responsible for the payment of intrastate access charges. In August 2003 the Alabama PSC opened a proceeding to consider that request. Initial comments were filed October 31, 2003, reply comments were filed December 2, 2003, and the matter is under review.

California - On September 30, 2003, the California Public Utilities Commission ("CPUC") asked six VoIP providers, including Vonage and Net2Phone, to apply by October 22, 2003 for the same license that landline phone companies need to operate in California. In response to that request, all six providers sent letters to the CPUC arguing that their VoIP services are exempt from state telephone regulations because they provide interstate information services that are not subject to the CPUC's jurisdiction. The CPUC then held a VoIP Forum on November 13, 2003 and has considered opening a formal inquiry into VoIP service regulation. The decision to open such proceedings has recently been at least temporarily delayed at the request of the lead commissioner based on her assessment that California should conduct any proceeding after the FCC has established national policy.[20]

Colorado – The Colorado Public Utilities Commission (the "Colorado PUC") opened a docket to determine the appropriate regulatory treatment of VoIP in May, 2003. The Colorado PUC closed the docket in January 2004, based in part on the "legal uncertainty of whether a state may regulate VoIP services," concluding that "the most prudent course is to take no action with respect to VoIP pending FCC action."[21]

Florida - The Florida legislature in 2003 passed, and the Governor signed, legislation stating "[that] the provision of voice-over-the-Internet protocol (VOIP) free of unnecessary regulation, regardless of provider, is in the public interest." The law also specifically excludes VoIP from the statutory definition of a "service" subject to regulation, although the question of whether VoIP-based services are subject to intrastate access charges remains under the jurisdiction of the Florida Public Service Commission.[22]

Minnesota - On August 13, 2003, the Minnesota Public Utilities Commission (the "Minnesota PUC") ruled that Vonage is offering a telecommunications service and

---

[20] *See* Ben Charny, *California to License VoIP Providers*, CNET News.com (Sep. 30, 2003), *available at* http://news.com.com/2100-7352-5084711.html?tag=guts_lh_7352. *See* also Ben Charny, *California eases up on Net phone rules*, CNET News.com (Jan. 5, 2004), *available at* http://news.com.com/2100-7352-5135188.html?tag=guts_lh_7352.

[21] Dkt. 03M – 220T, *See* p. 11 supra.

[22] *See The Tele-Competition, Innovation and Infrastructure Enhancement Act*, CS/SB 654 (FL, signed May 23, 2003).

required Vonage to seek a certificate, file a 911 plan and submit tariffs within 30 days. A US District Court granted Vonage's request to enjoin that decision on October 7, 2003 and the Minnesota PUC stayed its decision while it is enjoined. The district court ruled Vonage provides an "information service" not subject to Minnesota PUC jurisdiction. The Minnesota PUC requested the district court to amend its findings or to make its injunction temporary and to allow further investigation and discovery or grant a new trial. Oral argument took place on December 13, 2003. The District Court declined to amend any aspect of its order and concluded that a new trial was not necessary.[23]

Missouri - On September 12, 2003, while reserving its rights to argue for or benefit from any future regulatory determination relating to VoIP-based services, Time Warner Cable Information Services ("TWCIS") filed an application for authority to offer IP based voice services in Missouri. The parties to the resulting docket agreed that a general discussion of VoIP was not necessary but, although TWCIS had agreed to abide by existing Missouri telephone rules until the regulatory classification of VoIP is resolved, the parties disagreed about the characterization of the service TWCIS intends to offer and the related regulatory restrictions and obligations associated with that service. Separately, the Missouri Public Service Commission (the "Missouri PSC") sought comment from the Public Counsel as to whether it should open a generic proceeding to address regulatory issues surrounding VOIP services. The Missouri PSC subsequently chose not to open a generic proceeding, preferring instead to address issues in the context of the TWCIS application. A prehearing conference is scheduled for January 30, 2004. A proposed procedural schedule is to be filed by February 13, 2004.

---

[23] *See Matter of the Complaint of the Minnesota Department of Commerce Against Vonage Holding Corp. Regarding Lack of Authority to Operate in Minnesota*, Docket No. P-6214/C-03-108, Order Finding Jurisdiction and Requiring Compliance (rel. Sep. 11, 2003) (requiring Vonage to comply with all state laws pertaining to telephone service), enjoined, *Vonage Holdings Corp. v. Minnesota Public Utilities Comm'n*, No. 03-5287, slip op. at 22 (D. Minn. Oct. 16, 2003).

New York – The New York Public Service Commission (the "NYPSC") has ruled that VoIP service providers must pay access charges while preserving their right to be granted forbearance from regulation or to be alternately regulated based on any applicable decisions from the NYPSC or the FCC. The decision was based largely on the NYPSC's view that under the *Stevens Report* the company was operating as a phone-to-phone VoIP provider offering a "telecommunications service". Some parties have argued that the decision was based on a misreading of the report.

The NYPSC, pursuant to Frontier Telephone of Rochester's complaint against Vonage for providing telephone service without complying with state regulation, opened a generic investigation of VoIP issues. Initial comments were due October 31, 2003 and reply comments were due November 14, 2003. The matter is now under review.[24]

North Carolina – In May 2003, TWCIS applied for a certificate of public convenience and necessity to provide IP based voice services. The North Carolina Utilities Commission (the "NCUC") granted TWCIS its certificates in July 2003 and rejected efforts by the Alliance of North Carolina Independent Telephone Companies to address a number of issues in the context of the certification proceeding. At the time, BellSouth also sought a generic proceeding to address VoIP issues. The Commission determined that no such proceeding was necessary at that time.

Ohio – The Public Utilities Commission of Ohio (the "PUCO") opened a generic investigation in April 2003 to examine how VoIP services are provided, and the form and level of regulation that should apply to those services. Answers to PUCO questionnaires were filed in May, 2003; initial comments were filed on June 13, 2003

---

[24] *See, e.g., Complaint of Frontier Telephone of Rochester Against U.S. DataNet Corporation Concerning Alleged Refusal to Pay Intrastate Access Charges*, No. 01-C-1191 (N.Y. Pub. Serv. Comm'n May 31, 2002) (subjecting VoIP service to access charges, but preserving US DataNet's right to be granted forbearance from regulation or to be alternately regulated based on any applicable decisions from the NYPSC or the FCC); *Complaint of Frontier Telephone of Rochester Against Vonage Holding Corp. Concerning Provision of Local Exchange and Inter-Exchange Telephone Service in New York State in Violation of the Public Service Law*, No. 03-C-1285, Notice Requesting Comment (N.Y. Pub. Serv. Comm'n Oct. 9, 2003) (initiating a similar proceeding involving Vonage).

and reply comments were filed July 7, 2003. Since that time TWCIS has applied for, and has received from the PUCO, authority to provide service, contingent on the outcome of the generic investigation. TWCIS's application requested authority to provide IP voice services targeting the residential market using VoIP. TWCIS also requested waivers of various rules with which it found difficult to comply for its bundled service offering (in particular, offering stand-alone local service). The PUCO's decision granted waivers contingent on the outcome of the open investigation into whether VoIP technology should be regulated as a telephone service.[25] Since then, Cincinnati Bell, the Ohio Telecommunications Association, and SBC-Ohio filed applications for rehearing of TWCIS' application.

Pennsylvania – In May 2003 the Pennsylvania Public Utility Commission (the "Pennsylvania PUC") opened a generic investigation into VoIP and it is effectively forbearing from regulating those services pending the outcome of that investigation.[26]

Texas – In August 2003, TWCIS filed for a certificate of authority to provide IP based voice services in Texas. Several parties, including the Texas Coalition of Cities ("TCOC") attempted to intervene. In particular TCOC raised issues regarding the classification and jurisdictional status of the services proposed by TWCIS, and how compensation for rights-of-way would be administered for those services. The Texas Public Utility Commission (the "Texas PUC") denied intervention for all parties and it granted TWCIS' application on December 12, 2003.

Wisconsin - On September 11, 2003, the Wisconsin Public Service Commission (the "Wisconsin PSC") sent letters to VoIP providers 8x8, Vonage, and Delta 3 seeking information on the specific services being offered by those entities in Wisconsin. The PSC's letters stated that such entities were not permitted to provide resold intrastate services in Wisconsin without certification and that any customer bills for intrastate

---

25 *See* Public Utilities Commission of Ohio (Case 03-581-TP-ACE).

26 *Investigation into Voice over Internet Protocol as a Jurisdictional Service*, M-00031707 (May 5, 2003).

services were void and not collectible. [27] The providers filed responses which are under review.

V. NCTA's Approach: Balancing Responsibilities and Rights

Much of the discussion about VoIP services has focused on whether they should be classified as "information services," "telecommunications services," or another type of service. The assumption seems to be that VoIP service offerings first need to be assigned to a preexisting regulatory "box," from which a variety of regulatory consequences will flow. It is usually assumed that classification of a VoIP service as a "telecommunications service" means that it will be subject to a wide range of traditional Title II requirements, and that classification of a VoIP service as an "information service" means that it will be entirely unregulated. As discussed later in this paper, we believe neither assumption is correct.

Rather than focusing on this regulatory classification issue, NCTA suggests that policymakers focus on the responsibilities and rights that are appropriate for new entrant competitors offering VoIP services, whether they do so through their own facilities or over the facilities of others. The cable industry believes that VoIP service providers that meet the four-prong test described above must assume certain fundamental regulatory responsibilities, including consumer protections of general applicability, assistance to law enforcement, and public safety obligations. The industry also believes that in order to provide service, VoIP providers—particularly those operating their own facilities—must be accorded certain rights. The regulatory classification under which this set of responsibilities and rights is established is

---

[27] *See 8x8 Announces Receipt of Notification from Public Service Commission of Wisconsin*, 8x8 Press Release (Sep. 12, 2003), *available at* http://www.8x8.com/news_events/releases/2003/pr091203.asp.html.

important, though ultimately less important than those responsibilities and rights being established in a minimally regulatory framework.

VI. The Responsibilities and Rights of VoIP Providers

VoIP service providers, particularly those who build infrastructure that enables delivery of these services in competition with established local exchange carriers, must not be subject to unnecessary regulation, nor should they be disadvantaged as compared to VoIP providers who build no facilities. The strong presumption should be that regulations designed for legacy telephone service should *not* apply to VoIP services unless they are essential to meet the key public health, safety, and other crucial responsibilities described below, even if regulators determine they are necessary for customers of incumbent telephone utilities who may use VoIP technologies in substitution for legacy POTS services. Experience has shown, time and again, that the best way to encourage new and innovative technologies and to secure the resulting public benefits is to ensure that only the most vital regulations apply – and even then, that those vital regulations be adapted to the characteristics of the new technology.

This approach would encourage innovation, conserve regulatory resources, derive the greatest public benefits and provide the certainty in the marketplace that investors need in order to support the deployment of facilities-based VoIP services. The alternative – presuming that legacy regulations *do* apply, unless expressly found not to apply – is a recipe for doubt and delay. Few, if any, competitive communications technologies have ever achieved widespread market acceptance where government has followed that path; policymakers should be careful to avoid it here.

The set of responsibilities to which providers of services meeting the four-prong test should adhere may be broken into several categories: public health and safety; universal service; intercarrier compensation; and consumer protections of general applicability.

Public Health and Safety

Providers of VoIP services meeting the four-prong test should have the following responsibilities, implemented in a manner appropriate to the technology:[28]

- The obligation to cooperate with law enforcement, including compliance with the principles of CALEA based upon an IP-specific standard endorsed by an industry body.
- The obligation to provide consumers access to 911/E911 capabilities and to collect and remit funding for state or municipal 911/E911 systems. (In turn, statutory and other liability limitations for the provision of 911/E911 services should also apply.)[29]
- The obligation to make services available to disabled consumers, in a manner consistent with Section 255 of the 1996 Act, and to collect funding for state and federal TRS systems.[30]

Universal Service

In addition, regulators should expect VoIP services that make use of NANP resources to ultimately contribute to federal and state universal service programs on a par with other contributors. The principle of universal service – ensuring that affordable telephone service is available to high-cost areas and low-income users – has long been

---

[28] The FCC has ruled, for example, that, while facilities used solely for the provision of information services are not subject to CALEA, facilities used to provide both telecommunications and information services are subject to the requirements of the Act. *See Communications Assistance for Law Enforcement Act*, Second Report and Order, 15 FCC Rcd. 7105 at ¶¶ 12, 27 (1999). However, for both CALEA and 911/E911, some adjustments may need to be taken into account related to the specific features and capabilities of VoIP services.

[29] As with all service providers that offer 911/E911 capabilities, VoIP service providers should be protected by statutory and other limitations on liability pertaining to the provision of 911/E911 services.

[30] These rules have already been extended beyond the conventional range of Title II-type services, and the same considerations may apply to VoIP service. *See Implementation of Sections 255 and 251(A)(2) of the Communications Act of 1934, as Amended by the Telecommunications Act of 1996*, 16 FCC Rcd. 6417 at ¶ 8 (1999).

a cornerstone of communications policy. The 1996 Act codified principles of universal service and extended them to schools, libraries, and nonprofit rural health care providers.[31] Cable companies that offer telecommunications services subject to assessment currently pay into the fund.

At some point, VoIP services that make use of NANP resources should also pay into the fund. It would be premature to impose such an obligation, however, without resolution of several critical issues related to universal service, which the FCC is examining.[32] Among these issues is the question of whether the federal universal service fund is properly sized and funded.

It is critical that policymakers recognize the need to modify the current universal service contribution mechanism, particularly with respect to VoIP services.[33] Under the current contribution mechanism, assessments are based on *interstate telecommunications* revenues. Applying this mechanism to VoIP service would be fraught with difficulty for several reasons. First, because most consumer VoIP services today are offered without regard to interstate and intrastate distinctions, arbitrary judgments would be required as to which portion of VoIP service revenue is interstate and which is intrastate. Second, because the regulatory classification of VoIP service has not been determined, an arbitrary judgment would be required as to what portion of VoIP revenue is *telecommunications* revenue.

The best solution to this problem would be the adoption of a numbers-based contribution mechanism.[34] Any service which makes use of NANP resources would be

---

[31] *See* 47 U.S.C. § 254.

[32] In addition to the assessment methodology, other major unresolved issues include determining how high-cost support is computed; designating "eligible telecommunications carriers"; and reviewing the operations of the schools and libraries program (which the FCC had initially planned to conduct as part of a comprehensive universal service review in 2001, but which has not yet been initiated).

[33] See *Federal-State Joint Board on Universal Service, Report and Order and Second Further Notice of Proposed Rulemaking,* CC Docket No. 96-45, rel. Dec. 13, 2002 ("Second Further Notice").

[34] *See* Reply Comments of the National Cable & Telecommunications Association in *Second Further Notice*, April 18, 2003.

assessed on a per-number basis (special access and private line services would be assessed in a manner which results in a contribution approximately equal to that of today).[35] This is also consistent with the four-prong test previously described. Under such a system there would be no need to distinguish, for universal service purposes, between various types of VoIP offerings. *e.g.*, a voice service with the potential to substitute for a POTS line vs. a gaming service with a voice component. VoIP services that use telephone numbers would be assessed; those that do not use telephone numbers would not. At the same time, VoIP providers must be afforded nondiscriminatory access to universal service support. Any other approach would fail the competitive neutrality principle for universal service and discriminate against otherwise eligible providers based on technology.

Intercarrier Compensation

Similar considerations apply to intercarrier compensation rules. The issue here is not whether the rules should or should not apply but how to reconcile the many different rules – and different prices – that apply to exchanges of traffic.[36] Those differences, in turn, dictate not only different prices per unit of traffic, but also which party pays.[37] The FCC has a proceeding under way to resolve these issues.[38] When

---

35 See *AT&T Oct. 22 Ex Parte; Ad Hoc Oct. 3 Ex Parte* in *Federal-State Joint Board on Universal Service*, CC Docket 96-45, Further Notice of Proposed Rulemaking and Report and Order, FCC 02-43, rel. Feb. 26, 2002 ("Contribution Methodology Further Notice").

36 Today, the exchange of traffic is governed by a hodgepodge of different rules depending, for example, on whether an ILEC is exchanging traffic with a neighboring ILEC, a CLEC, an interexchange carrier ("IXC"), a CMRS provider, or an information service provider, and also depending on whether the traffic is deemed to be "intrastate" or "interstate."

37 For example, an ILEC handing off a call to a CLEC is required to pay that CLEC, but when an ILEC hands off a call to an IXC, the ILEC *receives*, rather than pays, compensation.

38 *See Developing a Unified Intercarrier Compensation Regime*, Notice of Proposed Rulemaking, 16 FCC Rcd. 9610 (2001).

that proceeding is concluded and the system has been rationalized, the new rules should apply to VoIP-based services that utilize the PSTN as well.[39]

Consumer Protection

In addition, generally applicable consumer protection rules that apply to all businesses should apply to VoIP service providers. These include such requirements as "do not call" and "do not mail." By contrast, as explained below, requirements that were developed to protect consumers from the monopoly utility in a single-provider environment are unnecessary and inappropriate.

Inappropriate Legacy Utility Requirements

VoIP services provided in competition with incumbent utility phone services should *not* be subject to legacy utility requirements designed largely in a monopoly environment. Most such requirements date from the era of a single provider of phone service and are inappropriate for competitors that offer alternatives to the incumbent providers. Legacy utility requirements all impose substantial burdens, none of which are justified in the case of competitive facilities-based VoIP services. The provider-subscriber relationship would be better served by consumer protection rules of general applicability, including appropriate disclosure requirements of any limitations of nonessential utility requirements, rather than the full panoply of detailed and cumbersome requirements applied to some public utility providers. In particular, a number of legacy requirements relate to billing, payment, credit and collection and quality of service standards. For example, many states have rules dictating the format and content of customer bills; rules regarding permitted forms of payment, the allocation of partial payments, and in-person payment obligations; and rules regarding call center

[39] This proposal presupposes that equitable rules will be established for all classes of entities that exchange traffic. If classification as an interexchange carrier, Internet service provider, etc. triggers differing compensation regimes, then the problems of arbitrage and gamesmanship will be perpetuated. Under the current rules various classes of entities may have an economic incentive to

metrics, installation intervals, and service establishment requirements. This is but a partial list of utility provider requirements that typical competitive entrants should not face.

As competition increases, marketplace forces, rather than prescriptive rules, can address these issues much more effectively – subject to informing potential customers, so they can make judgments about the service. For instance, because of the industry-wide trend (spurred by consumer demand) towards bundled products and services, various legacy utility mandates such as equal access, tariffing, and dialing parity are simply inappropriate, and particularly so where VoIP services are bundled with services which are not subject to such requirements.[40] VoIP providers may, however, choose to adopt them on a voluntary basis. But, any unnecessary rules will increase costs for VoIP providers and deter investment, delay deployment, and slow the growth of these promising new services. Regulators should make a comprehensive effort to identify and eliminate all such unnecessary rules. This will be an essential element of a successful VoIP policy.

Rights of VoIP Providers

Just as VoIP service providers meeting the four-prong test must accept certain responsibilities, such providers require certain rights. These rights must be available to the provider irrespective of whether the provider's service is ultimately determined to be an "information service," a "telecommunications service," or another type of service. Additionally, granting these rights should not influence the regulatory classification of the VoIP service.

---

deliver traffic in an uneconomic or inefficient fashion in order to avoid high intercarrier compensation rates.

40 Notions of "equal access" may be inapplicable to (or prevent the offering of) innovative service packages that give a customer a fixed quantity of usage for a set monthly price, and/or where there is no price differentiation between local and long distance calls.

These rights include, but are not limited to: (1) the right to interconnect and efficiently exchange traffic and control signaling with both IP and PSTN entities on a peer-to-peer basis;[41] (2) the right to obtain telephone numbers, including numbers secured through number portability, to assign those numbers to VoIP customers and to have them published in the telephone directories; (3) the right to access the facilities and resources necessary to provide VoIP customers with full and efficient 911/E911 services (*e.g.*, interconnection to incumbent utility E911 selective router switches, and Master Street Address Guide and Automatic Location Identification database uploads); (4) the right to be compensated fairly for terminating traffic delivered from other entities, in accordance with the results of an industry-wide review of payments for traffic termination and origination that specifically addresses VoIP service;[42] and, (5) the right to non-discriminatory access to universal service support.

Policymakers must also ensure that facilities-based VoIP service providers have the right to use rights-of-way, including pole attachments, ducts, and conduits. Moreover, VoIP services delivered by cable operators wiil normally be conveyed over pre-existing facilities already attached to poles, located in underground conduits or crossing rights-of-way. Accordingly, policymakers must ensure that cable operators are not subject to additional or incremental assessments and fees when they change the pattern of signaling in their pre-existing physical transmission paths to add VoIP services to their existing video and Internet offerings. In addition to unnecessarily and unjustifiably burdening cable operators' VoIP services, such fees and assessments would put cable operators at a competitive disadvantage vis-à-vis incumbents who usually control such essential facilities, and non-facilities based providers of VoIP

[41] Including access to codes needed for network interconnection and traffic exchange with other providers and the PSTN, NPAC databases and capabilities, SS7 interconnection for call management between VoIP calls and the PSTN, and customer service records housed in ILEC/CLEC databases.

[42] This is an area where it would be sensible for a PUC to await FCC rulings on petitions pending before that body, rather than to make determinations applicable only to intrastate VoIP service traffic, or that might be out of harmony with what federal regulators ultimately require for interstate VoIP traffic.

services who utilize cable facilities to make their offerings available. In particular, higher pole rates should not be a barrier to entry for facilities-based VoIP providers.[43]

VII. Regulatory Restraint and Regulatory Classification

As noted, the cable industry's approach to a VoIP regulatory framework is to focus on the responsibilities and rights appropriate for providers meeting the aforementioned four-prong test, rather than focusing on the regulatory classification of those services. But those issues cannot be avoided. NCTA supports the view of FCC Chairman Michael Powell that VoIP services warrant a fresh assessment, from a highly deregulatory perspective. We agree that policymakers should, as Chairman Powell has stated; "build from a blank slate up as opposed to from the myriad of telecommunications regulations down. . . . [I]t is a nasty, entangled litigious exercise to start from a phone company world of regulation and work your way down this way, rather then to try to say, no, this is something new."[44]

Though complex, the challenge of developing an appropriate regulatory framework for new network applications is not entirely new to the FCC. The FCC's decision in the *Second Computer Inquiry (Computer II)*[45] to eliminate regulation for

---

[43] The FCC has statutory authority to establish an appropriate pole attachment rate for attachments by cable operators. Setting an appropriate rate would be an important part of creating a hospitable environment to encourage the deployment of facilities-based VoIP offerings. *See National Cable Telecommunications Association v. Gulf Power*, 534 U.S. 327 (2002).

[44] *See* Remarks of FCC Chairman Michael K. Powell at the Meeting of the Technology Advisory Council, at 2 (Oct. 20, 2003). *See also* Powell VoIP Forum Remarks at 1 ("As one who believes unflinchingly in maintaining an Internet free from government regulation, I believe that IP-based services such as VoIP should evolve in a regulation-free zone. No regulator, either federal or state, should tread into this area without an absolutely compelling justification for doing so."). The results of this exercise may also produce insights that could also be applied to traditional circuit-switched, facilities-based CLEC services. Clearly, *all* CLECs lack market power, and sound public policy (as well as the dictates of the 1996 Act) commands that *all* unnecessary regulation of telecommunications services should be avoided.

[46] *See Amendment of Section 64.702 of the Commission's Rules and Regulations*, Final Decision, 77 FCC 2d 384 at ¶ 84 (1980) ("*Computer II*"), *aff'd sum nom. Computer & Comm. Ind. Ass'n v. FCC*, 693 F.2d 198 (1982) (subsequent history omitted). It was *Computer II* that prevented federal or state

"enhanced services" and customer premises equipment led to investment and innovation that reverberates more than twenty years later. Likewise, the Commission's decision to forbear from entry and exit regulation as well as tariffing requirements for CMRS[46] produced similarly salutary results.[47]

Conversely, application of the full panoply of traditional telecommunications regulation would impede deployment of facilities-based VoIP services.[48] Only in an environment in which the burdens of regulation are kept to a reasonable minimum will potential VoIP providers be in a position to deploy sustainable facilities-based VoIP services quickly and to their full potential. Such an environment enjoys broad governmental and industry support.[49] In this regard, Congress has directed the FCC

---

regulation of interstate information services. See 77 FCC 2d 384 at ¶ 7. *Computer II* also ensured the deregulation and competitive provision of customer premises equipment ("CPE"). *See id* at ¶ 9.

46 *See Implementation of Sections 3(N) and 332 of the Communications Act, Regulatory Treatment of Mobile Service*, 9 FCC Rcd. 1411 at ¶¶ 173-182 (1994) (subsequent history omitted) (forbearing from many Title II requirements, stating that "Congress and the Commission have determined that the public inherently benefits from the promotion of competition among the carriers that results from market-based pricing for their services"). *See also Petition of the People of the State of California and the Public Utilities Commission of the State of California to Retain Regulatory Authority over Intrastate Cellular Service Rates*, Report and Order, 10 FCC Rcd 7486 at ¶¶ 96-97 (1995) (denying a California PUC petition to extend state regulatory authority over CMRS services). Recognizing that wireless services operate without regard to state boundaries, Congress also preempted state and local rate and entry regulation of CMRS. 47 U.S.C. 332(c)(3).

47 *See Annual Report and Analysis of Competitive Market Conditions with Respect to Commercial Mobile Services*, Eighth Report, 18 FCC Rcd. 14783 at ¶ 57 (2003) (noting the results of the deregulatory environment created for wireless carriers by the FCC: "Continued downward price trends, the continued expansion of mobile networks into new and existing markets, high rates of investment, and churn rates of about 30 percent . . . demonstrate a high level of competition for mobile telephone consumers"). This report also noted that wireless subscribership increased in 2002 to over 141 million users in the U.S., *see id.* at ¶ 59, a tenfold increase in less than a decade.

48 While it is clear that unnecessary regulation would create a significant business problem for circuit-switched CLECs, the case against excessive Title II regulation of VoIP services is even more compelling. Circuit-switched telephony is an existing service, using proven technologies. By contrast, VoIP service uses nascent technologies that have yet to be deployed on any significant commercial scale, and which could present a host of as-yet-undetermined financial, technical, and operational challenges. As noted above, the development of a minimally regulated environment for VoIP services ought to provide a basis for revisiting -- and reducing -- the regulatory requirements that apply to traditional circuit-switched, facilities-based CLEC services.

49 Numerous policy leaders (including many in the FCC and in state government), industry representatives and others have recognized the importance of limiting regulation of facilities-based

and the state PUCs to "encourage the deployment on a reasonable and timely basis of advanced telecommunications" by "utilizing . . . regulatory forbearance . . . [and] other regulating methods that remove barriers to investment."[50]

For the reasons detailed above, public policy strongly and unquestionably favors a pro-competitive, deregulatory approach to facilities-based VoIP services. Fortunately, federal law and FCC precedents are largely consistent on this point. However, state laws and regulation are varied; as described above, states have taken widely differing approaches to VoIP – ranging from minimal regulation in states such as Florida to attempts to apply full common carrier service regulation in states such as Minnesota. NCTA's view is that state regulation of VoIP services should be consistent with FCC regulatory treatment. State consistency with federal regulation is important because an Internet-based service has an interstate (even global) reach; 51 different approaches would make it difficult to develop VoIP service.

And federal leadership for the states will also prevent a legal logjam where one state regulatory regime, if appealed, becomes law in that region of the country while the rest of the nation comes to follow the federal scheme. This anomaly is not theoretical. One panel of the U.S. Court of Appeals for the 9th Circuit ruled that its earlier decision on the regulatory classification of cable modem service – reached before the FCC had made its own regulatory determination – continued to govern. That prior determination held, regardless of the analysis made by the FCC and despite the usual deference

---

VoIP services. FCC Chairman Michael Powell and FCC Commissioners Martin and Abernathy have called for regulatory restraint with respect to VoIP services. *See, e.g.*, Cable Monitor, *FCC and NTIA Call for Regulatory Protection for VoIP*, Aug. 26, 2002. Similar – if not more strongly deregulatory -- statements were made by multiple FCC Commissioners at the FCC's Dec. 1, 2003 VoIP Forum. Acting NTIA Administrator Michael Gallagher is reported to have said that "any regulation of VoIP should be 'minimalist and narrowly tailored' to meet public interest goals" and that excessive regulation could drive providers overseas. *See* Communications Daily, *Powell Sees FCC Focusing on Discrete Issues on VoIP*, at 2 (Dec. 2, 2003) ("*CommDaily Report on VoIP Forum*").

50 Pub. L. No. 104-104 § 706, 110 Stat. 56 (1996); *see also* 47 U.S.C. § 157(a) (establishing federal policy of encouraging the provision of new technologies and services to the public).

owed to expert agencies over just these sorts of policy questions.[51] A premature state decision could lead to a similar unfortunate result in the VoIP policy context.

In considering how to proceed under the Act, both state and federal regulators would do well to consider the "nascent services doctrine,"[52] articulated by FCC Commissioner Kathleen Abernathy. It is a set of principles, which, while not a legal mandate, is instructive for policymakers.

This doctrine recommends that regulators exercise restraint when dealing with new technologies and services and to reevaluate the need for any regulation of those technologies and services as they evolve. Such restraint would facilitate the development of facilities-based VoIP services that compete with the established telephone companies without the burden of anachronistic regulations and would promote the goal of enhancing facilities-based local telephone competition.[53]

The doctrine further suggests that once new facilities-based competitors demonstrate their viability, policymakers and regulators reexamine the overall

---

[51] *See Brand X Internet Services v. FCC* 345 F.3d 1120 (9th Cir. 2003) (per curiam); *AT&T Corp v. City of Portland* 216 F.3d 871 (9th Cir. 2000). *See also Chevron U.S.A., Inc. v. Natural Resources Defense Council*, 467 U.S. 837 (1984).

[52] *The Nascent Services Doctrine*, Remarks of FCC Commissioner Kathleen Q. Abernathy Before the Federal Communications Bar Association, New York Chapter (Jul. 11, 2002), available at http://www.fcc.gov/Speeches/Abernathy/2002/spkqa217.html.

[53] In a sense, this is what the Commission did in the *Stevens Report* where, by essentially deciding not to address the regulatory classification of VoIP services, it allowed for five years of technology development, service experimentation, and capital investment. *See Stevens Report*, 13 FCC Rcd. 11501 at ¶¶ 86-93 (1998). Similarly, in the AT&T/TCI Merger and in the first Report to Congress under § 706, the FCC declined to interfere with emerging high-speed cable Internet services, thereby fostering the massive investment that today makes broadband service available to 80 percent of American homes. *See Applications for Consent to Transfer the Control of Licenses and Section 214 Authorizations from Tele-Communications, Inc., Transferor, to AT&T Corp., Transferee*, Memorandum Opinion and Order, 14 FCC Rcd. 3160 at ¶ 94 (1999); *Inquiry Concerning the Deployment of Advanced Telecommunications Capability to All Americans in a Reasonable and Timely Fashion, and Possible Steps to Accelerate Such Deployment Pursuant to Section 706 of the Telecommunications Act of 1996*, First Report, 14 FCC Rcd. 2398 at ¶ 106 (1999); National Cable & Telecommunications Association, *Cable & Telecommunications Overview, Mid Year 2003* at 10, at http://www.ncta.com/pdf_files/Mid'03Overview.pdf (stating that 85 million of approximately 106 million U.S. homes had access to cable broadband service at the end of 2002).

regulatory scheme applicable to incumbent providers in the marketplace to assess whether existing regulations applicable to incumbents should be modified. If appropriate, regulatory schemes over time would be harmonized, but with much less regulation than previously, reflecting the effects of competition.

The focus of the "nascent services doctrine" is not on establishing the appropriate regulatory classification (*i.e.*, whether a VoIP service is a "telecommunications service," an "information service," or another type of service), but on how best to allow both facilities-based and non facilities-based VoIP services to develop naturally in the marketplace in response to consumer demand and technological innovation. Applying this doctrine, regulators would avoid those regulations that will unnecessarily hinder the evolution and growth of a new service, and ultimately lessen all regulation as competitive circumstances warrant.

While adherence to the principles of the nascent services doctrine is a worthwhile goal, policymakers must follow such principles within the context of an appropriate statutory framework. Based on the appropriate set of responsibilities and rights, as articulated above, VoIP providers need an approach which either begins with Title I and layers on responsibilities and rights, or begins with Title II and forbears significantly from a number of responsibilities -- effectively a Title "1.5."

More specifically, the FCC and the states can secure a reasonable and minimally regulatory environment for VoIP services through classification of VoIP applications as "information services" under Title I of the Communications Act. An alternative but potentially more problematic approach would be to use the FCC's "forbearance" and preemption powers under Title II to minimize regulation. Each path is discussed briefly below.

Title I Regulatory Approach

The designation of certain VoIP services as information services – and the use of Title I ancillary authority to impose only those regulations that are essential to helping regulators meet key public health, safety, and other responsibilities – is the primary way in which policymakers could minimize burdens on these emerging services. Since *Computer II*, designation of a service as a Title I information service has meant that it is deregulated, in the sense that it is not subject to common carrier regulation by federal or state regulators.[54] Even a Title I service, however, can be regulated under the FCC's "ancillary authority," but only in furtherance of specific statutory objectives.[55]

A pure Title I approach may be particularly well suited to certain forms of VoIP services that provide capabilities and features that make them markedly different from conventional phone services. Examples of such services may include video phone, voice chat, and video chat services. Depending on their characteristics, however, even VoIP services that more closely resemble conventional telephone offerings may well meet the definitions of an information service. Specifically, VoIP services could be designed in ways that easily satisfy the statutory definition, *i.e.*, "the offering of a capability for generating, acquiring, storing, transforming, processing, retrieving, utilizing, or making available information via telecommunications." [56] They could even more easily be designed to satisfy the enhanced service definition of *Computer II*, *i.e.*, services "which employ computer processing applications that act on the format, code, protocol or similar aspects of the subscriber's transmitted information; provide the subscriber additional, different, or restructured information; or involve subscriber interaction with stored information."[57]

As noted above, classification of a service as an information service does not necessarily mean that it should be exempt from all regulation. The FCC retains "ancillary authority" under Title I to adopt those regulations that are reasonably

---

[54] *See Computer II*, 77 FCC 2d 384 at ¶ 84 (1980).

[55] *See People of State of Cal. v. F.C.C.* , 905 F.2d 1217, 1241 at n.35 (9th Cir.) (1990).

[56] 47 U.S.C. § 3(20).

[57] 47 C.F.R. § 64.702(a).

necessary to advance explicit statutory objectives.[58] We have already outlined the social responsibilities appropriate for VoIP providers whose service meets the four-prong test described above, where those responsibilities are associated with certain rights. Significantly, classification of VoIP service under Title I does not mean that those rights could not be conferred on VoIP providers. For example, it is likely that the Commission could order local exchange carriers to interconnect with Title I VoIP providers or even provide unbundled network elements. Prior to 1996, using its Title II authority over local exchange carriers, the FCC ordered the Bells to interconnect with information service providers in the *Expanded Interconnection*[59] and *Computer III*[60] proceedings.

After enactment of the 1996 Act, the Commission sought comment on whether those requirements were still valid and appropriate.[61] As of now, the requirements remain in effect. Nevertheless, it is an open issue whether the 1996 Act, by establishing specific interconnection and unbundling duties of local exchange carriers that are owed only to providers of telecommunications services, precludes the Commission from imposing the same or similar duties on carriers for the benefit of VoIP providers.

NCTA emphasizes that the rights set forth in Section VI supra are critical to any VoIP regulatory regime under Title I, including interconnection, eligibility to receive universal support and participation in a sustainable intercarrier compensation regime.

---

[58] *See People of State of Cal. v. F.C.C.*, 905 F.2d 1217, 1241 at n.35 (9th Cir.) (1990).

[59] *Expanded Interconnection Order*, 7 FCC Rcd 7369, ¶ 65 (1992).

[60] *Computer III Phase I Order*, 104 FCC2d 958 ¶ 113 (1986); *Computer III Further Remand Proceeding*, 10 FCC Rcd 8630 ¶ ¶ 18-19 (1995).

[61] *Computer III Further Remand Proceeding*, Notice of Proposed Rulemaking, 13 FCC Rcd 6040 (1998); Request to Refresh the Record, 16 FCC Rcd 5363 (2001).

Regulatory Forbearance and Preemption Under Title II

The FCC has (and PUCs may have) considerable authority to decide that even "telecommunications services" need not be subject to various requirements under Title II of the Communications Act. For example, the FCC's *Competitive Carrier* rulemaking, which scales regulatory responsibilities according to the presence or absence of market power associated with a particular service, allows the FCC to eliminate regulations for entities or classes of providers that have low market shares and no potential to acquire and to wield market power.[62] Obviously, facilities-based VoIP service providers, newly entering the market, who compete against dominant 100-year-old telephone service providers, will have little or no ability to engage in the abuses that full common carrier regulation is designed to prevent.

Building upon the principles of the FCC's *Competitive Carrier* decision, Congress in the 1996 Act created a mechanism of regulatory restraint that extends not only to FCC-made rules but also to statutory provisions. Under Section 10 of the 1996 Act, the FCC is empowered and required to eliminate *any statutory or regulatory requirement* that applies to any telecommunications service or telecommunications service provider if: (1) the requirement is unnecessary to prevent unfair and unjust charges and practices, (2) enforcement of that requirement is not needed to protect consumers, and (3) forbearance would otherwise serve the public interest.[63] VoIP services offered by new entrants, especially in their initial phases, are ripe for Section 10 forbearance.

---

[62] *See Policy and Rules Concerning Rates for Competitive Common Carrier Services and Facilities Authorizations Therefor*, First Report and Order, 85 FCC 2d 1 at ¶ 4 (1980) ("*Competitive Carrier*"). *See also Policy and Rules Concerning Rates for Competitive Common Carrier Services and Facilities Authorizations Therefor*, Fifth Report and Order, 98 FCC 2d 1191 at ¶¶ 19-23 (1984) (forbearing from most regulation of nationwide common carrier digital transmission networks ("DEMS"), holding that forbearance will help promote the entry and expansion of DEMS by relieving carriers of the costs and delay of required tariff filings and will help promote competition).

[63] *See* 47 U.S.C. §160(a). Some parties have sought Section 10 forbearance under the 1996 Act. *See e.g.*, CTIA *Petition for Forbearance from Section 310(d) Regarding Non-Substantial Assignments of Wireless Licenses and Transfers of Control Involving Telecommunications Carriers*, 11 C.R. 61 (1998), *Forbearance From Applying Provisions of the Communications Act to Wireless Telecommunications Carriers*, 21 C.R. 802 (2000).

Such regulatory restraint is essential to promote investment, innovation, and widespread deployment.

The FCC followed this line of reasoning in its cable modem Declaratory Ruling and NPRM. There it said that "to the extent cable modem service is classified as a telecommunications service [in the 9th circuit] ... forbearance would be in the public interest because cable modem service is still in its early stages; supply and demand are still evolving; and several rival[s] ... are still developing. For these same reasons [the Commission] tentatively conclude[s] that enforcement of Title II provisions and common carrier regulation is not necessary for the protection of consumers or to ensure that rates are just and reasonable and not unjustly or unreasonably discriminatory. As such, [the Commission] believe[s] that forbearance from the requirements of Title II and common carrier regulation is appropriate in this circumstance."[64]

There are several observations about "forbearance" worth noting. First, this approach ordinarily presumes that Title II requirements and rules apply in the first instance, and then eliminates them one (or a few) at a time. A more flexible and deregulatory approach might couple the notion of forbearance with the "nascent services doctrine" so as to identify only the Title II requirements appropriate to VoIP and forbear from the rest in accordance with the standards of Section 10. Such an approach would ensure that VoIP services are never subject to the full panoply of Title II-type regulations, but rather are subject, from the outset, only to those regulatory obligations that have been affirmatively determined to be necessary.

Second, forbearance can be slow; at the federal level, telecommunications service providers must apply for forbearance, either individually or as a class, and the FCC may take up to 15 months (during which time regulation continues) before a final

---

[64] *See Inquiry Concerning High Speed Access to the Internet over Cable and Other Facilities, Appropriate Regulatory Treatment for Broadband Access to the Internet over Cable Facilities,* FCC GN Dkt, Nos. 00-185, 02-52 Declaratory Ruling and Notice of Proposed Rulemaking, at ¶ 95 (rel. Mar. 15, 2002)

decision is rendered.[65] This problem can be solved if the FCC takes action promptly, through its contemplated NPRM[66] and through other appropriate steps to provide a measure of regulatory certainty for VoIP services.

Third, FCC forbearance standing alone operates only to curtail interstate regulation but does nothing to address excessive and inconsistent *intrastate* phone regulations.[67] Two solutions to this problem are apparent. One is for PUCs to embrace a light-handed regulatory approach and ensure that any state regulation of VoIP services is consistent with FCC regulatory treatment. Failing that, the other solution is for the FCC to use its preemption powers to constrain state action. Indeed, a determination under Title I that VoIP is an interstate information service would preempt states by definition. If VoIP is classified as a telecommunications service under Title II, then Section 253 requires the FCC to preempt state laws, regulations, and rules that prohibit or have the effect of prohibiting any entity from providing such services.[68] More broadly, the FCC has preexisting preemption powers, resident in Sections 1, 2, and 4(i) of the 1996 Act, to preempt state regulations that impede the provision of interstate communications services.

* * *

Given the range of possible paths to a suitably deregulatory regime, there appears to be every reason for federal and state policymakers to embrace a minimally regulatory regime for VoIP services, so that vast numbers of residential consumers will enjoy the benefits of competition, new and exciting services will be introduced, and new

---

[65] *See* 47 U.S.C. §160(a).

[66] *See* XChange, *FCC to Open Proceedings on VoIP Regulation,* Nov. 7, 2003 (citing a letter from FCC Chairman Michael Powell to U.S. Senator Ron Wyden, in which Powell stated that: "Over the course of the next year, after full public comment and thoughtful consideration of the record, the FCC plans to follow up . . . [an] NPRM with a report and order on the VoIP issues raised in the proceeding.").

[67] But note that a number of state public utility commissions also operate under laws that allow for the exercise of regulatory forbearance.

[68] *See* 47 U.S.C. § 253.

investment and jobs will be stimulated. Only a regulatory framework that is minimally burdensome can create the right incentives and a favorable climate in which service providers can invest, innovate, and deploy VoIP services.

Conclusion

Cable's massive investment since the 1996 Act has enabled the industry to offer a host of new services. These services include high-speed Internet access, digital cable, HDTV and video-on demand. Several cable companies also have substantial circuit-switched telephony operations. VoIP, however, is more than just the next new application. The cable industry believes that VoIP technology will permit cable companies to provide innovative, high-value facilities-based residential local phone services at competitive prices across the U.S. Such services, especially offered by facilities-based providers like cable competitors, hold the promise of breaking the logjam that has long denied consumers the benefits of real and sustainable competition and choices for local telephone service. While cable companies are excited about the potential benefits that can result from the widespread availability of VoIP services, a broad rollout cannot be assured unless a (de)regulatory framework applies to these services.

If policymakers affirmatively embrace and promote VoIP services, and keep them free of unnecessary and inconsistent regulation, the result will be to attract additional investment and propel rapid and ubiquitous deployment. This is the lesson to be drawn from the broadband explosion since the 1996 Act: pro-competitive, deregulatory policies work as nearly 18 million cable modem customers bear witness. Conversely, public benefits will inevitably be reduced and delayed if unnecessarily restrictive regulations from the monopoly telephone era are applied. The choice is clear.

NCTA
NATIONAL CABLE & TELECOMMUNICATIONS ASSOCIATION
1724 Massachusetts Avenue NW
Washington, DC 20036
TELEPHONE: (202) 775-3622
WEBSITE: www.ncta.com

U.S. Department of Justice
Office of Legislative Affairs

Office of the Assistant Attorney General

Washington, D.C. 20530

July 23, 2004

The Honorable Chris Cannon
Subcommittee on Commercial
and Administrative Law
Committee on the Judiciary
U.S. House of Representatives
Washington, DC 20515

Dear Mr. Chairman:

We appreciate this opportunity to present the Department of Justice's views on telephone service that uses the Internet Protocol ("VoIP"). How Congress will treat this service will profoundly impact law enforcement's ability to protect communities across the nation from the harms inflicted by drug trafficking, organized crime, and terrorism and fundamentally to protect the national security of the United States. It is important that public safety and national security concerns be carefully considered as part of any legislation addressing advanced communications technology.

I. VoIP Is an Important Technology.

First, we want the Subcommittee to know that the Department of Justice believes that telephone service that uses the Internet Protocol has the potential to provide tremendous benefits. We are hopeful that this form of telephone service will cost less, provide better service, and include exciting new features. The Administration strongly supports the rapid and widespread rollout of new technologies such as VoIP that make America more productive. As with all new technologies, the Department of Justice celebrates the benefits that VoIP promises, while at the same time working vigorously to protect our country and citizens against its misuse.

II. Electronic Surveillance is a Critical Tool For Protecting Public Safety and National Security.

As part of our work to protect our country and citizens, we want to underscore how very important it is that no type of telephone service becomes a haven for criminals, terrorists, and spies. Access to telephone service, regardless of how it is transmitted, is a highly valuable tool for investigating and even preventing crime. Not only is electronic surveillance one of the most effective tools to combat crimes such as terrorism, espionage, and organized crime, but it is often the only effective tool.

Any criminal conspiracy requires communication in order to operate. Today, these communications often do not occur in person, where law enforcement could observe a meeting taking place -- could see people physically coming and going. Instead, criminals do what many of us do, they use the telephone. Telephones allow criminals to coordinate their activities and allow organizers and kingpins to keep their hands clean of the most sordid criminal conduct.

Federal and state prosecutors often note the importance of evidence gathered through electronic surveillance in obtaining arrests and convictions. Last year alone, 3,674 people were arrested based on evidence obtained through wiretaps. Over the past ten years, wiretap evidence has led to over 54,000 arrests. That is up to 54,000 criminals that might have escaped justice had court-ordered electronic surveillance not been available.

Electronic surveillance not only provides otherwise unobtainable evidence of criminal activity, but it also helps prevent crimes and save lives. For instance, in his 1994 testimony, former Director of the Federal Bureau of Investigation ("FBI"), Louis Freeh, described how electronic surveillance led to prevention of terrorist attacks such as a planned rocket attack against an FBI field office and an attack on a nuclear power facility.

Electronic surveillance is also a critical law enforcement tool to identify and dismantle organized criminal networks, including major national and international drug cartels. Last year, a wiretap in California led to seizures of literally thousands of tons of illegal drugs and millions of dollars. Another wiretap investigation led to over one hundred arrests, as law enforcement dismantled an international drug distribution ring that was responsible for vast quantities of heroin and cocaine coming into the United States from Columbia. Electronic surveillance has allowed us to take cocaine, heroin, methamphetamine, and many other dangerous drugs off our streets and away from our children.

Because electronic surveillance is such an effective law enforcement tool, criminals go to great lengths to shield their telephone communications. One tactic they employ is to use a wide array of communication devices, trying to isolate the damage done if a particular means of communicating is compromised. For instance, a recent Drug Enforcement Administration investigation revealed a Miami drug trafficker who is known to have used 20 different cellular phones in a three-month period in an attempt to evade law enforcement surveillance.

What is more, we know that when it becomes known that law enforcement has difficulty detecting communications over a particular technology, criminals quickly migrate to that technology. While we obviously cannot go into detail on this point, suffice it to say that criminals do not want to be caught, and they are quick to take advantage of any gap in our ability to detect and disrupt their criminal activities.

If criminals could use new technologies to avoid detection, they could use these technologies to coordinate terrorist attacks, to distribute drugs throughout the United States, and to pass along national security secrets to our enemies. If the criminals were successful, we would learn about these plots only after terrible damage had been done, or in some cases not at all. Put simply, law enforcement cannot effectively protect the public and enforce the laws in today's world without electronic surveillance.

III. Because Electronic Surveillance Is Such A Powerful Tool, It Is Rightfully Subject To Equally Powerful Limits On Its Use.

While electronic surveillance is a necessary tool, we are mindful that it is also a very powerful tool, which has serious implications for the privacy of citizens. As such, we only use electronic surveillance as a tool of last resort, and even then we adhere to strict limitations on its use.

First, the U.S. Constitution places important parameters on our use of electronic surveillance. Under the Fourth Amendment, the government must demonstrate probable cause to a neutral magistrate before obtaining a warrant for a search, arrest, or other significant intrusion on privacy.

Congress and the courts have also provided statutory limits beyond those required by the Constitution. For instance, law enforcement must obtain a "trap and trace" or "pen register" court order in order to obtain information identifying who is sending or receiving communications to or from a particular suspect, even though not required under the Constitution. *See* 18 U.S.C. 3121 et. seq.

The Wiretap Act, 18 U.S.C. §§ 2510-22 ("Title III"), places an even higher burden on the real-time interception of the content of wire communications. The Senate Report on Title III stated explicitly that the legislation "has as its dual purpose (1) protecting the privacy of wire and oral communications and (2) delineating on a uniform basis the circumstances and conditions under which the interception of wire and oral communications may be authorized." Senate Committee on the Judiciary, Omnibus Crime Control and Safe Streets Act of 1967, S. Rep. No. 1097, 90th Cong., 2d Sess. (1968) at 66. Accordingly, under Title III, in order to obtain a court order to capture communications as they occur, the government must show that normal investigative techniques for obtaining information about a serious felony offense have been or are likely to be inadequate or are too dangerous, and that any interception will be conducted so as to ensure that the intrusion is minimized.

Even beyond the limits placed by the Constitution and the Congress, the Department of Justice has its own internal procedures to provide still more safeguards. For example, the Office of Enforcement Operations ("OEO") in the Criminal Division of the Department reviews each proposed Title III application to ensure that the request for interception satisfies the protections of the Fourth Amendment and complies with applicable statutes and regulations. Even if OEO recommends authorizing a request, the application cannot go to a court without approval by a Deputy Assistant Attorney General or higher-level official in the Department. The fact that not a single application for electronic surveillance under Title III was rejected by a federal court in all of 2003 is a testament to the vigilance and care the Department takes when asking for this authority.

If the Department of Justice approves a federal Title III request, it still must, of course, be submitted to and approved by a court of proper jurisdiction. The court will evaluate the application under the Fourth Amendment and using the familiar standards of Title III. By statute, for example, the application to the court must show, through sworn affidavit, why the intercept is necessary as opposed to other less-intrusive investigative techniques. The application must also provide additional detail, including whether there have been previous interceptions of communications of the target, the identity of the target (if known), the nature and location of the communications facilities, and a description of the type of communications sought and the offenses to which the communications relate. By statute and internal Department regulation, the interception may last no longer than 30 days without an extension by the court. All intercepted communications are sealed by the court, further protecting privacy.

Often courts also impose their own safeguards. For example, many federal courts require that the investigators provide periodic reports to the court setting forth information such as the number of communications intercepted, the steps taken to minimize irrelevant traffic, and whether the interceptions have provided information relevant to the criminal investigation. The court may, of course, terminate the interception at any time.

The remedies for improperly intercepting communications in violation of Title III or the Electronic Communications Privacy Act ("ECPA") can include criminal sanctions, civil liability, and, for law enforcement agents, adverse employment action. For violations of the Fourth Amendment, of course, the remedy of suppression is also available.

All of these requirements and procedures ensure that electronic surveillance is only used when absolutely necessary to detect and prosecute serious criminal violations. It is a tool of last resort reserved for only the worst offenses against our civil society. It is done with the approval and oversight of the courts, and done in ways as narrowly tailored as possible to the investigation of specific individuals for specific criminal conduct. Further, if it is misused, there are serious consequences.

IV. CALEA is Critical to Implementing Court Orders Authorizing Electronic Surveillance.

While electronic surveillance is a critical tool for law enforcement, it is not always easy to implement, and it is becoming even more difficult. In the past, when law enforcement agencies conducted court-authorized electronic surveillance, they were able to go to one provider and access a "local loop" that allowed a single location for the collection of content and related dialing information for all communications with the subject's telephone number. However, it has been a long time since all that was required to implement a court order for electronic surveillance was a call to Ma Bell and a set of alligator clips.

Today, communications are transmitted over many different wires and cables and over a myriad of frequencies through the air. These communications are provided by many different companies who use many different protocols. It is because of both the breadth of services and the technical complexity of features associated with each one that law enforcement relies on the designers to assist in providing interception capability for the select cases where a court has ordered such interception.

The Congress has already recognized this problem and taken decisive action to prevent public safety and national security from being imperiled as a result of the digital communications revolution. In 1994, Congress "concluded that there is sufficient evidence justifying legislative action that new and emerging telecommunications technologies pose problems for law enforcement." In response, you prudently passed the Communications Assistance for Law Enforcement Act ("CALEA").

In enacting CALEA, you made clear that the purpose of the statute "is to preserve the government's ability, pursuant to court order or other lawful authorization, to intercept communications involving advanced technologies such as digital or wireless transmission modes, or features and services such as call forwarding, speed dialing and conference calling, while protecting the privacy of communications and without impeding the introduction of new technologies, features and services." As the legislative history makes clear, "the bill seeks to balance three key policies: (1) to preserve a narrowly focused capability for law enforcement agencies to carry out properly authorized intercepts; (2) to protect privacy in the face of increasingly powerful and personally revealing technologies; and (3) to avoid impeding the development of new communications services and technologies."

In crafting this solution, Congress wisely did not limit CALEA's scope to just one particular technology, service, or suite of features, but rather set in place a structure that anticipated and provided for a vast array of technological advances. As the then Director of the FBI testified in support of the legislation, it was

> intended to stand the test of time and overcome the shortcomings of the 1970 amendment. It is specifically designed to deal intelligently and comprehensively with current and emerging telecommunications technologies and to preclude the need for much more restrictive and more costly legislation in five or ten years when court-authorized interceptions would no longer be possible due to further technology advances. Any legislation that would limit its application to technological impediments on a piecemeal basis would be disastrous. Piecemeal legislation which deals only with current problems or some of the problems would result in common carriers fully deploying new technologies which would impede electronic surveillance and which would cause the government to return to Congress repeatedly.

Hearing on Police Access to Advanced Communications Systems Before the Senate Subcommittee on Technology and the Law of the Committee on the Judiciary and the House Subcommittee on Civil and Constitutional Rights of the Committee on the Judiciary (statement of Louis J. Freeh, Director of the Federal Bureau of Investigation) ("Freeh CALEA Testimony").

Thus, Congress has already recognized the importance of ensuring that, as advanced telephone service technologies develop, they must have the technical ability to implement court orders for surveillance.

Now, ten years later, we must not back away from the important principles behind CALEA. If anything, it is even more critical today than in 1994 (when CALEA was enacted) that advances in communications technology not provide a haven for criminal activity and an undetectable means of death and destruction.

It is important to be very clear here - we ask only that Congress not undermine current capabilities to implement court orders and conduct critical law enforcement activities. CALEA is about the practical necessity of implementing existing lawful authority, not expanding it. Congress said so itself, noting in the legislative history to CALEA that "[s]ince 1968, the law of this nation has authorized law enforcement agencies to conduct wiretaps pursuant to court order. That authority extends to voice, data, fax, E-mail and any other form of electronic communication. The bill will not expand that authority." Nothing in CALEA gives law enforcement the authority to conduct any surveillance. It is only after all of the comprehensive regulatory, statutory, and Constitutional protections described above have been complied with that CALEA comes into play and ensures that the order of the court can be carried out.

In fact, CALEA explicitly and intentionally protects privacy in very important ways. As the House of Representatives explained in its report on CALEA, "the bill further protects privacy by requiring the systems of telecommunications carriers to protect communications not authorized to be intercepted." It does this in two ways. First, CALEA requires that providers be able to separate out the communications of just the subscriber for whom law enforcement has an order to intercept communications. This provision benefits both efficiency and privacy. Second, CALEA requires that a service provider be able to separate out call-identifying information from the content of communications. This protects the call content from law enforcement access where law enforcement only has legal grounds to obtain the call-identifying information.

V. The Application of CALEA to Advanced Telecommunications Technologies Is at Issue in Proceedings Before the Federal Communications Commission.

The Federal Communications Commission ("FCC") recently issued its Notice of Proposed Rulemaking concerning the appropriate treatment of IP-enabled services, including telephone service that uses the Internet Protocol. Hundreds of parties have submitted their thoughtful consideration of the issue, including the Department of Justice.

With regard to CALEA in particular, the Department of Justice has petitioned the Commission for an expedited rulemaking to clarify which services and entities are subject to CALEA. We expressed our view that broadband access and telephony service providers are "telecommunications carriers" under CALEA and, therefore, must be capable of implementing court orders for surveillance.

In both the IP-enabled services and CALEA proceedings at the FCC, the Department of Justice has made the following points:

(1) that public safety and national security will be compromised unless court orders for electronic surveillance can be implemented in a timely fashion by providers;

(2) that assistance requirements should apply to every service provider that provides switching or transmission, regardless of the technologies they employ; and

(3) that if any particular technology is singled out for a special exemption from these requirements, that technology will quickly attract criminals and create a hole in law enforcement's ability to protect the public and the national security.[1]
The CALEA proceedings in particular are creating a compelling record regarding the drastic consequences if we were to fail to provide law enforcement the tools it needs to protect public safety. Thus far, dozens of state and local law enforcement entities - from New York to Los Angeles and dozens of places in between - have filed comments at the FCC emphasizing the critical need for these tools and the dire consequences of failure. It is not surprising that so many police chiefs and district attorneys came out in strong support of the Petition, because state and local governments account for almost three-fifths of all wiretap applications. As the National Association of District Attorneys expressed so well in their comments to the FCC in the CALEA Rulemaking proceeding:

> For over a decade we have been pleading for the tools and the laws we need to protect the people in our communities. We will never know whether we could have prevented the tragic

[1] As noted in the Department's filings at the FCC, the determination that a service provider is subject to CALEA does not mean that the carrier must then be subject to economic regulation traditionally applied to common carriers under the Communications Act of 1934, as amended. For purposes of CALEA, VoIP should not be defined as an exempt "information service." Such a finding under CALEA does not preclude deregulation under the Communications Act.

> consequences of September 11th had we had the investigative tools we have been asking for since 1992. We only know that we will need every advantage to prevent such a tragedy from ever occurring again.

Comments of the National Association of District Attorneys, In the Matter of Joint Petition for Rulemaking to Resolve Various Outstanding Issues Concerning the Implementation of the Communications Assistance for Law Enforcement Act, RM-10865, at 2.

We are also pleased that a number of the Commissioners have already publicly acknowledged the need to preserve law enforcement access to telephone service that uses the Internet Protocol.[2] Chairman Powell was unequivocal in his statement accompanying the recent IP-Enabled Services Notice of Proposed Rulemaking. He stated:

> CALEA requirements can and should apply to VoIP and other IP-enabled service providers, even if these services are "information services" for purposes of the Communications Act. Nothing in today's proceeding should be read to suggest that law enforcement agencies should not have the access to communications infrastructure that they need to protect our nation. On the contrary, all IP-enabled services should consider the needs of law enforcement as they continue to develop innovative technologies.

Statement of Chairman Michael K. Powell, In the Matter of IP-Enabled Services: Notice of Proposed Rulemaking, FCC 04-28.

Further, many responsible members of the communications industry have agreed with the Department that their assistance is critical to public safety and national security. One industry association put it simply: "American citizens should be assured that communications companies are providing appropriate help to law enforcement." Comments of the United States Telecommunications Association, In the Matter of IP-Enabled Services: Notice of Proposed Rulemaking, FCC 04-28, at 36-37.

There is one aspect of the Department's position in the CALEA proceedings before the FCC that is important to clarify to avoid misunderstanding. Law enforcement

[2] See Statements of Chairman Michael K. Powell and Commissioners Kathleen Q. Abernathy, Kevin J. Martin, and Johnathan S. Adelstein, In the Matter of IP-Enabled Services: Notice of Proposed Rulemaking, FCC 04-28.

does not seek the power to dictate how the Internet should be engineered or the power to veto the deployment of new telecommunications services. CALEA specifically states that it "does not authorize any law enforcement agency or officer to require any specific design . . . ." 47 U.S.C. 1001(b)(1)(A). Nor does CALEA authorize law enforcement "to prohibit the adoption of any equipment, facility, service, or feature . . . ." 47 U.S.C. 1002(b)(1)(B). As law enforcement requested, Congress made the providers' obligations under CALEA generic by design. The then Director of the FBI could not have been more clear on this point, when he testified in support of the CALEA legislation in 1994:

> The Government purposely eschewed setting any technical standards because it does not desire to 'dictate' particular technological solutions. It is the Government's position that each common carrier is best positioned and qualified to determine how it will meet the requirements in the most cost-effective way.

Freeh CALEA Testimony.

Law enforcement cannot – nor do we seek to – dictate to any carrier how best to design its service or what services it can or cannot offer. We only ask that any service comply with the law in order not to imperil public safety and national security.

We also want to take this opportunity to address the contention that applying CALEA to VoIP services will stifle the development of VoIP technologies or drive its providers out of the country. This argument relies on the premise that CALEA compliance will be either impossible or unreasonably burdensome. As some have already told the FCC, compliance is not only possible, but solutions are already "available today in the commercial marketplace at reasonable prices." See., e.g., Comments of Verisign, Inc. on Joint Petition for CALEA Rulemaking at 7. Moreover, it will be much more cost-effective to design in these capabilities than to try to engineer post-hoc solutions once an order is served and time to comply with the order could become a matter of life and death. Finally, consistent with CALEA's protection of both innovation and public safety, CALEA has built in safety valves that allow providers to petition the FCC for relief from CALEA obligations if compliance with CALEA is not "reasonably achievable." 47 U.S.C. § 1006. In determining whether compliance is "reasonably achievable," CALEA directs the FCC to consider, among other things, the effects on customer rates, financial resources of the provider, and U.S. policy to encourage new technology.

The Department of Justice appreciates your support as we continue with the difficult work of protecting our nation and enforcing our laws during times of rapid

The Department of Justice appreciates your support as we continue with the difficult work of protecting our nation and enforcing our laws during times of rapid technological change. We are concerned that the failure to preserve the application of CALEA to new technologies, such as VoIP, could create a safe haven for criminal activity. It is very important that, in taking action regarding telephone service that uses the Internet Protocol, Congress carefully consider implications to public safety and national security.

If we may be of additional assistance, we trust that you will not hesitate to call upon us. The Office of Management and Budget has advised that there is no objection from the standpoint of the Administration's program to the presentation of this report.

Sincerely,

William E. Moschella
William E. Moschella
Assistant Attorney General

cc: The Honorable Melvin L. Watt
Ranking Minority Member

11

Mr. CANNON. Before I begin with the witnesses' introductions, interested parties will likewise have 5 days to submit written statements.

I am now pleased to introduce today's hearing witnesses. Our first witness is Robert Pepper, chief of policy development at the Federal Communications Commission. In this capacity, Mr. Pepper has served as the direct advisor to FCC Chairman Michael Powell on long-term policy planning. He is also the co-chair of the FCC's Internet Policy Working Group and has primary responsibility for developing the Commission's overall relationship with the financial community. Prior to his fulfilling his current appointment, since March 2003, Mr. Pepper was chief of the FCC's Office of Plans and

Policy beginning in 1989. Mr. Pepper has published and lectured widely on telecommunications policy issues. He is a graduate of the University of Wisconsin-Madison, where he received his doctoral degree.

Our next witness is John Langhauser, vice-president, law, and chief counsel to the Consumer Services Group of AT&T Corporation. Mr. Langhauser joined AT&T in 1982 and has held legal positions in the State government affairs, antitrust litigation, international business services, Federal regulatory and public policy groups. Prior to joining AT&T, he was a litigator with the firm of Dewey Ballantine in New York. Mr. Langhauser graduated cum laude from Harvard Law School and summa cum laude from the State University of New York at Plattsburgh.

Our next witness is Mr. Stephen Cordi, deputy comptroller for the Maryland Comptroller of the Treasury. Mr. Cordi has served in this capacity since 1994 and has primary responsibility for tax administration. He is also the immediate past president of the Federation of Tax Administrators. Mr. Cordi was the first director of the Compliance Division of the Maryland Comptroller following its creation in 1993. For 13 years prior to this appointment, he was the director of the Maryland Sales and Use Tax Division. Mr. Cordi first entered State service in 1974 as special assistant to the Attorney General for the comptroller. An attorney and certified public account, he is a graduate of Haverford College and Georgetown University Law Center.

Our final witness is Mr. James Kirkland, general counsel and senior vice-president of Covad Communications. Mr. Kirkland is responsible for overseeing all of Covad's legal issues relating to regulatory and legal affairs, corporate governance and employment and finance. Prior to joining Covad, Mr. Kirkland served as general counsel and senior vice-president of Spectrum Development for the privately-held Clearwire Technologies, Inc., a broadband Internet service provider based in Dallas, Texas. Before joining Clearwire, Mr. Kirkland spent 17 years with Mintz, Levin, Cohen, Ferris, Glosky and Papeo, P.C., located here in Washington, D.C., where he specialized in communications law. Mr. Kirkland holds a bachelor's degree from Georgetown University and a law degree with honors from Harvard Law School.

I extend to each of you my warm regards and appreciation for your willingness to participate in today's hearing. In light of the fact that your written statements will be included in the record, I request that you limit your oral remarks to 5 minutes. Accordingly, please feel free to summarize or highlight the salient points of your testimony. I can assure you that you will have more time to explain particular points thereafter.

You will note that we have a lighting system that starts with a green light. After 4 minutes, it turns to a yellow light, and then, at 5 minutes, it turns to a red light. It is my habit to tap the gavel at 5 minutes. We would appreciate it if you would finish up your thoughts within that time. We don't expect you to just cut off. We are actually anxious to understand what you think is important for us to understand, but that is a time frame that will actually help us move through the hearing.

After all the witnesses have presented their remarks, the Subcommittee Members in the order that they arrive, and I suspect that is just one other Member, will be permitted to ask questions of the witnesses subject to the 5-minute time limit. Pursuant to the directive of the Chairman of the Judiciary Committee, I ask the witnesses to please stand and raise your right hand to take the oath.

[Witnesses sworn.]

Mr. CANNON. The record should reflect that each of the witnesses answered in the affirmative.

You may be seated, and Mr. Pepper, would you now proceed with your testimony?

TESTIMONY OF ROBERT PEPPER, PH.D., CHIEF, POLICY DEVELOPMENT, OFFICE OF STRATEGIC PLANNING AND POLICY ANALYSIS, FEDERAL COMMUNICATIONS COMMISSION

Mr. PEPPER. Good morning, Mr. Chairman and distinguished Members of the Subcommittee.

It is my pleasure to come before you this morning to talk about voice over Internet protocol or V-o-I-P or VoIP. VoIP services and applications are dramatically expanding beyond the limited functionality of traditional telephone voice service and at the same time challenging the traditional economic and regulatory structures that have governed the traditional telephone industry for more than a century.

Saying that VoIP is just another way to make a phone call is much like saying that Ebay is just another way to have a garage sale. This ignores the fact, obviously, that ecommerce and the Internet have fundamentally changed the way we compare products and prices, transact business and the way service providers compete for and relate to consumers. VoIP is best understood as bringing this dynamic to the market for voice communications. The traditional network delivered voice over brilliantly-designed, dedicated and centrally-managed network. Whoever owned the pipe into your home owned the customer.

On the Internet, however, the voice application and, in fact, all applications are separated from the physical transmission network. They ride over that network but are agnostic as to who provides it. Thus, anyone who can attach a server to the Internet can allow two, three, four, 100 people to talk to one another. Voice is becoming little more than one application of many over a multiuse, digital broadband network, less like standalone phone service and more like a free or almost free add-on to something else you can buy from multiple sources.

Indeed, the majority of voice-over-IP applications, including voice instant messaging and talking to players of live interactive games like X-box, look nothing at all like traditional telephone service. These are fundamental changes in an industry that has been regulated for almost a century on the assumption that all providers are monopolies, protected by an elaborate regulatory regime in which they use dedicated narrowband networks. It would be irrational for regulators to ignore these changes and automatically apply legacy regulation without first seriously examining whether it is relevant.

History provides two excellent examples of a better way: cell phones and the Internet. These technologies were largely freed of common carrier regulation, notwithstanding long, hard-fought battles to impose it. Today, the American consumer and the American economy are far better off for having steered a deregulatory course. These two industries grew from reaching just a handful of customers to bringing substantial benefits to tens of millions in the absence of any significant common carrier regulation.

The Commission has begun examining VoIP issues in this light in a notice of proposed rulemaking regarding IP-enabled services as well as in specific petitions. The Commission began its reexamination of VoIP because development of this promising technology might very well be hampered by unjustified, conflicting and burdensome regulatory requirements that could result as different State commissions and courts begin to address the area.

In this environment, the Commission cannot simply assume that inaction will create an environment that encourages innovation, investment and competition. In response to the NPRM, the Commission received over 150 comments and 86 reply comments from a very wide variety of parties. The Commission already has issued two orders resolving petitions for declaratory ruling, one filed by Pulver.com and the other by AT&T. In addition, the Commission is considering VoIP-related petitions from Vonage, Level 3, SBC and Inflection.

The Commission is also considering questions related to voice over IP and its universal service contribution, intercarrier compensation and our upcoming CALEA proceeding. The Commission's decisions in this area will have the farthest-reaching consequences of anything the Commission currently is considering. What is at stake is nothing less than the future of electronic communications for future generations.

The Commission, however, is constrained by the Act, which divides the world into regulated telecom services and unregulated information services. When dealing with revolutionary new technologies, we need to start from the perspective of how to best create the world we all want to live in rather than applying tired regulations soon to be rendered obsolete. While the Commission has some ability to fine-tune treatment of new technologies, given its discretion and flexibility granted to it by Congress, the Commission's latitude is limited by the Act.

If you believe that VoIP and other new technologies are transforming the telecom market in ways that cry out for new regulatory approaches, you need to consider whether the tools the Commission has today are adequate for that task. In the meantime, the Commission is moving forward with its work, and guidance and leadership from Congress is crucial to the success of our process.

Mr. Chairman, on behalf of the FCC, I want to thank you for calling this hearing, and we look forward to working with you and other Members on these issues.

[The prepared statement of Mr. Pepper follows:]

PREPARED STATEMENT OF ROBERT PEPPER

Good morning, Mr. Chairman and distinguished members of the Subcommittee. It is my pleasure to come before you today to discuss services and applications that use voice over Internet Protocol ("VoIP"), and the status of our examination of VoIP at the Federal Communications Commission (the "FCC" or the "Commission").

I. THE IMPORTANCE OF VOIP

Voice over Internet Protocol services and applications are dramatically expanding beyond the limited functionality of traditional voice telephone service and, at the same time, challenging the traditional economic and regulatory structures that have governed the traditional telephone industry for more than a century.

The FCC has pending before it a number of proceedings initiated by petitioners about VoIP, and has initiated a broad examination of issues related to VoIP, as well as other Internet Protocol (IP) based services. As an introduction to the status of these proceedings, it is helpful to discuss why the emergence of VoIP raises important issues, why the Commission, as indicated in the IP-Enabled Services Notice of Proposed Rulemaking ("IP-Enabled Services Proceeding"), is examining the best way to establish a minimally regulated environment for VoIP, and why prompt action to clarify the regulatory regime applicable to VoIP is crucial to the future of electronic communications and America's place as the leading innovator in the field.

A. VoIP is Changing the Nature and Business of Voice Communication

VoIP is seen by some as simply a new technology for transmitting a traditional voice telephone call. This purely functional view, sometimes referred to as the "if it quacks like a duck, it's a duck" argument, is short-sighted for two reasons.

VoIP Technology is Radically Different From Traditional Voice Telephony. First, the functional view ignores the fact that VoIP technology is merely an application that rides over the public Internet, or over dedicated data networks, just like any other application. On these public or private data networks the bitstream created by a VoIP application is no different than any other bitstream on that data network—it can be incorporated into other bitstreams, modified or enhanced by simply changing server or client software. Thus, voice can now be easily combined with data and video in ways that cannot be done over the traditional network. Adding enhancements to voice, or incorporating voice to other applications, is merely a question of adding a new feature in the next software release. With VoIP, consumers can easily change their service selections or add function and enhanced features simply by logging on to their VoIP application provider's website, or by choosing a new provider with more attractive features. And, by the way, the majority of voice over IP applications look nothing like traditional plain old telephone service. Some of these include voice instant messaging or the ability to talk to opponents while playing a game across the Internet on XBox Live.

VoIP is a Radically Different Way of Doing Business. The second reason why a purely functional approach is short-sighted is that it is a new way of doing business. As my colleague, Jeffrey Carlisle has noted, saying VoIP is just another way to make a phone call is like saying that Amazon.com is simply a new way to sell books, without any broader consequences for markets or consumer behavior. E-commerce is much more than that. It changed the market for books, and everything else, by opening a truly worldwide market to any retailer who could attach a server to the Internet, or any individual who could open an E-Bay account.

Similarly, VoIP changes the business of telecommunications by allowing data networks to carry voice communications at comparable levels of quality to the traditional circuit-switched network, but to do so more flexibly and efficiently. VoIP changes the dynamics of the market for telecommunications services in three ways.

First, VoIP transforms voice from the primary service provided by a common carrier into just another application on the network. On traditional telephone networks, voice was delivered over a dedicated network that required a well-capitalized infrastructure and service provider that traditionally was a protected monopoly. In the future, the voice application—in fact, all applications—will be separated from the physical transmission network. Anyone can attach a server to the Internet, anywhere in the world, to allow two people—or three, four, five or a hundred—to talk to one another, just as anyone can connect a server to the Internet to provide email, file sharing, or other applications. The implications for how voice services are marketed and purchased are dramatic. No longer is the monopoly provider the gatekeeper for innovation. Rather, innovation in telecommunications can come from any entrepreneur, small company or enterprise that can connect to the network. This is the consequence of moving voice communications to the Internet, where intelligence is on the edge of the network instead of a tightly controlled core.

With these kinds of developments, saying that a VoIP application is merely another way of making a phone call is like saying that the automobile is just another way of going someplace in your horse and buggy. VoIP means that voice may no longer be a dedicated service for which consumers pay a separate monthly bill. VoIP may be part of your wireless phone service, as it already is with many push-to-talk services; it may be bundled together with video and data service that you buy from your cable, telephone, satellite or power company; or you may buy it from dozens of providers over the Internet; or you may simply have it as part of a software package that you buy for some other purpose. Most likely, you will acquire it in all of these different ways. When VoIP separates the voice application from the physical network, the question will no longer be whether consumers will benefit from competition in the voice market. Clearly, they will. Rather, competition in voice will no longer be an issue, because voice will become an almost free add-on to something else you buy from multiple sources.

In this respect it is useful to compare the evolving voice market to email. Email appears to be "free," but email application providers thrive in a market where intense competition drives innovation. Advances in email provided by Google, Yahoo! and Hotmail become headline news. Consumers can acquire email applications from their ISP, select web-based mail from third parties supported by advertising, outsource mail services, or operate email servers on their own networks. In the same way, consumers will benefit from a market for voice applications thriving with competition, innovation and choices suited to their needs at significantly reduced costs—but with significant rewards for agile and smart companies capable of delivering the best service.

The second way VoIP is changing telecommunications markets is that it accelerates the migration to all digital, multiuse broadband infrastructures. Whatever the benefits of separating the voice application from a dedicated infrastructure, there still need to be companies capable of building and maintaining the digital infrastructure over which applications ride. For most, if not all, markets in the United States, infrastructure will no longer be the monopoly domain of the traditional telephone network. Instead, an entire range of broadband technologies, including DSL, cable modem, licensed and unlicensed wireless broadband, Ultra Wide Band, satellites and broadband over power line will provide connectivity. When networks provide transmission, and are not tied to a single application like voice or video, networks become highly substitutable and competition increases dramatically, resulting in significant benefits for consumers. Additionally, the offering of demand-creating applications such as VoIP promotes deployment and adoption of broadband facilities, which in turn promotes further development of VoIP and other Internet applications. Thus, applications and broadband create a virtuous cycle that will result in significant benefits for American consumers and the American economy as a whole.

The third way VoIP changes telecommunications markets is that it internationalizes voice communications. Just like many other applications provided over the Internet, it doesn't matter where the provider is located—a server providing a VoIP application could be down the street, or in the next state, or it could be in Ukraine, the UK, India, or, as is currently the case with Skype, in Estonia. A voice application provided on a server located in a foreign country, with the customer in the U.S. using nothing more than software downloaded from the Internet and purchasing a broadband connection from a third party, looks very different from the service provided by traditional phone companies. This fundamental shift in how the voice application is provided has obvious implications for what regulations, if any, are imposed on VoIP providers and who decided and/or enforces any regulation. Federal or state regulators need to recognize that it may be very difficult to enforce requirements and unwarranted burdensome regulation will place VoIP providers in this country at a competitive disadvantage to VoIP providers located in relatively less regulated countries, and that, if providers are driven abroad, we will lose desirable jobs in the high technology sector.

Much of what I have described is a look into the reasonably foreseeable future. But VoIP is already changing the market's dynamics, even though it has not yet become ubiquitous. In 1998, VoIP generated less than 0.2% of the world's international voice traffic. In 2002, VoIP generated 10.4%, and, in 2003, is estimated to have generated 12.8%. Recently, Cablevision announced that it would provide a bundled package of digital cable TV, high speed Internet, and unlimited local and long distance calling for $90. If you consider what consumers pay for digital cable and broadband in the marketplace today, at this price, the voice service is essentially free. This is exactly what one would expect when voice, which uses relatively little bandwidth, is provided over a high bandwidth connection.

There are other indications that VoIP, while only gradually making its way into the public consciousness, is nevertheless growing at an increasing pace. A report released last month by the Pew Internet & American Life Project and the New Millennium Research Project estimates that approximately 14 million Americans have already made some sort of voice communication over the Internet. Skype, an Internet-based VoIP service that allows its members to speak to one another with crystal clarity for free over a peer-to-peer network connection, has been downloaded over 15 million times by users around the world.

B. Why Take Action Now?

The FCC has long relied on a policy of limiting regulatory intrusion on the Internet and applications provided over it. The Commission could have waited and raised the question of how VoIP is regulated at some point in the future, after it matured. At the end of 2003, incumbent local exchange carriers ("ILECs") and competitive local exchange carriers ("CLECs") served over 181 million access lines in the United States, and even at astronomical growth rates it will be some time before VoIP services and applications constitute a significant portion of the U.S. voice market. But there are two factors pressuring for Commission attention and, by implication, legislative action.

First, industry players are deploying these applications today, and are bringing their questions to the Commission. VoIP only started to become more widely used in the domestic market within the last several years. Thus, beginning in September 2002, a variety of companies from across the telecommunications industry—VoIP applications providers, ILECs, data companies and interexchange carriers ("IXCs")—filed petitions with the Commission seeking clarification regarding regulatory treatment of VoIP.1 The petitions filed over the last two years demonstrate the need for clarification and a measure of certainty on important regulatory questions, especially since it is uncertain how the FCC is going to rule in this very new environment.

Second, because of the important traditional role state public utility commissions play in regulating intrastate telecommunications, states have now begun to look at these questions, raising the possibility of differences among state regulatory regimes, and between various state and federal regulatory regimes. Some state commissions have decided to wait until this service further develops or until the FCC acts. But others have moved forward to examine VoIP, and some, such as Minnesota and New York, have already taken steps to classify VoIP applications as regulated telecommunications services. Federal courts in both states have stayed the effectiveness of these rulings. Nevertheless, companies offering VoIP are dealing today with multiple attempts to apply potentially inconsistent regulatory regimes, with the imminent prospect of more to come. This uncertainty and potentially conflicting regulatory regimes is an impossible position for companies wanting to provide VoIP service on a national basis.

It is not surprising, therefore, that while there is investment capital that would fuel even further innovation in this high tech area, there is hesitance to bring this capital to market while the regulatory regime remains unclear. While this might be said of any number of areas of telecommunications law, it is particularly true of VoIP, given that much of the innovation in the area is coming from small companies and entrepreneurs who are most vulnerable to shortages of investment capital. Therefore, the FCC has begun to examine this area not because it is looking for something to do, or because it is interested in any way in regulating the Internet. Rather, the FCC has begun to examine this area because of the demonstrated need for clarity because of the very real possibility that deployment of this new technology will be hampered by burdensome and conflicting regulatory requirements.

II. THE IP-ENABLED SERVICES PROCEEDING

On March 10, 2004, the Commission released its Notice of Proposed Rulemaking ("NPRM") on IP-Enabled Services, docket number 04–36 in order to address the need for the Commission to provide clarity to consumers, industry and the investment community. This NPRM asked commenters to tell the Commission how it could best craft a regime for VoIP and other IP enabled services that would encourage innovation and ensure that the benefits of this technology could reach consumers.

The NPRM discusses how VoIP will change how voice service is delivered to business and residential customers, and then asks whether the Commission can best serve the public interest by continuing its policy of minimal regulation of the Internet and applications provided over it. It asks for comment on how the Commission could determine whether a service using VoIP is a regulated telecommunications service or an unregulated information service under the 1996 Act: Should the Commission establish the line at the point where VoIP technology interfaces with the public switched telephone network? Should the Commission use a purely functional approach that makes the distinction based on whether the given service is a replacement for traditional telephony? Should the Commission use a test that examines whether the service substitutes for traditional telephony as determined by a traditional market analysis? Should the Commission instead adopt a layered approach, view VoIP purely as an application riding over a network, and thus regulate applications very lightly while applying a more stringent regime to facilities? And what impact should it have on the Commission's analysis that VoIP can be provided via peer-to-peer services that simply connect two users, as opposed to the centrally managed networks used by traditional service providers? In the case of traditional service providers, there is an entity to regulate that, presumably, has some control over and information about the calls routed over its network. In the peer-to-peer case, consumers communicate directly with one another, and aside from assisting in linking the participants, the provider of the peer-to-peer application may have little or no control over the call.

Related to the question of classification, the NPRM asks how the Commission might best achieve a minimally regulated environment. If classified as an information service, the service is nevertheless subject to the Commission's general jurisdiction to regulate all interstate and international communications by wire and radio. Alternatively, even if a service is classified as a telecommunications service, Congress has directed the Commission to forbear from enforcing its own regulations or the requirements of the statute if enforcement is not necessary to protect consumers, ensure against unjust, unreasonable or unreasonably discriminatory practices, or protect the public interest.

The NPRM goes on to solicit comment as to jurisdiction. It notes the Commission's recent order in response to a petition for declaratory ruling filed by Pulver.com regarding Free World Dialup—as described in the petition, a free peer-to-peer application facilitating voice communication between members of a closed group, which does not interconnect with the public switched telephone network. The Commission's Order, released on February 19, 2004, held that Free World Dialup was an information service subject to federal jurisdiction. The Pulver.com order further held that state regulation treating Free World Dialup like a regulated telecommunication service would most likely be preempted given the Commission's finding and an explicit Congressional policy against burdening the Internet with unnecessary federal and state regulation. The NPRM acknowledges that the Pulver.com Order only addressed one type of VoIP, and asked about the extent to which the reasoning in the case can be applied to other types, such as VoIP applications that interface with the public switched telephone network.

Having solicited comment on how the Commission should classify VoIP, and who should have jurisdiction as to whether to regulate VoIP, the NPRM then asks what regulations, if any, should apply, and develops an important distinction. The NPRM asks whether economic regulations such as entry, exit, tariff and accounting rules designed to protect against the power of a monopoly provider of services, with control over the bottleneck facility of the wire into the consumer's home, have any application in an environment where consumers can choose any number of applications providers, and use those applications over multiple networks. If technology has redressed the imbalance in power between customers and providers by lowering barriers to entry and allowing the consumer to choose his or her service provider, and change that choice easily, does this type of legacy economic common carrier regulation continue to have any relevance, at least as regards VoIP providers? Certainly, precedent indicates that where competitive choice is possible, lower regulatory burdens are justified. This has been the case with cellular providers, which are not subject to many of the common carrier requirements that might otherwise apply to them. It has also been the case with nondominant wireline providers. The NPRM solicits comment on these issues.

Traditional economic common carrier regulation is distinguished from requirements that can be characterized as social obligation regulation. These are requirements that, as a society, we have decided should apply broadly to any provider of voice services, as opposed to only those providers that have a dominant market position. Thus, even if a provider of voice is not dominant, we still believe that it is important that its customers have access to emergency services. Even if the market for voice services is changing in fundamental ways, it is still a basic goal of the Communications Act to ensure that all Americans have access to service at affordable prices. One might say that free voice service would achieve that goal. But if it is necessary to first purchase some form of broadband service, then it may be necessary to examine how we understand universal service and support for it may need to change over time. The social obligations raised in the NPRM and related proceedings include emergency service via the 911/E911 system, access to telecommunications by people with disabilities, universal service, and authorized law enforcement access to electronic communications—important societal goals that should not be compromised as the market changes. But the NPRM recognizes that the ways to achieve these goals are likely to change as the result of widespread VoIP adoption. Thus, while it makes clear these goals continue to be important, the NPRM also asks how the Commission can best achieve them in the new environment, acknowledging both the difficulties and opportunities presented by new technology.

III. COMMENTS ON THE IP-ENABLED SERVICES PROCEEDING

The response by the public to the NPRM has provided the Commission with a rich record, and features original and thought-provoking analyses of the issues. By May 28, 2004, the date for filing of initial comments, the Commission had received over 150 sets of comments. And, by last count, the Commission has received 86 reply comments by the July 14 filing date. These comments and replies have come to the Commission from a wide range of sources, indicating the broad interest this proceeding engenders not only among industry actors, but across American society as a whole. These sources include:

- multiple public utility commissions, and two organizations representing state commissioners, the Federation for Economically Rational Utility Policy and the National Association of Regulatory Utility Commissioners;
- county 911 administrators;
- the Department of Homeland Security and the Department of Justice;
- groups involved in studying and advocating public policy as it relates to high tech issues, such as the Electronic Frontier Foundation;
- public interest groups representing specific groups of consumers, such as AARP, the American Foundation for the Blind, Communication Service for the Deaf, the National Consumer League and the Ad Hoc Telecommunications Users Committee;
- trade groups representing the interests of telecommunications and high tech industries, including the Telecommunications Industry Association, CTIA, NCTA, the Information Technology Association of America, and the High Tech Broadband Coalition;
- cable TV providers, including Cablevision, Time Warner, and Comcast;
- wireless providers, including Nextel, Cingular, Ericsson, and T-Mobile; and
- Internet Service Providers;
- many well-known high technology companies such as Microsoft and Cisco;
- local exchange carriers, both incumbent and competitive, as well as their trade associations;
- rural telephone companies, as well as their trade associations; and
- numerous VoIP application providers, such as 8X8, Net2Phone, Skype, Pulver.com, Callipso, Dialpad, Vonage, and the Voice on the Net Coalition.

The commenting parties have, by and large, acknowledged the significant changes that VoIP technology will bring. They differ, however, as to the specific regulatory implications of that change.

A number of commenters, largely state commissions and rural incumbent local exchange carriers ("rural ILECs"), argue that if VoIP provides the functional equivalent of a voice call, then it should be regulated in the same way as traditional voice telephony. Others argue for a multi-factor test to determine whether a service should be regulated or not. For example, the National Cable Television Association argues that a VoIP application should be subject to the same regulation as telecommunications service providers if the following applies: (1) it makes use of 10 digit numbers under the North American Numbering Plan; (2) it is capable of receiving calls from the public switched telephone network at one or both ends of the call; and (3) it represents a possible replacement for traditional telephone service. However, NCTA also argues that if a service meeting all of these criteria also uses IP protocol between the service provider and the consumer, including use of an IP terminal adapter and/or IP-based telephone set, it should be subject to minimal regulation. Still others, such as AT&T, SBC, many of the high technology companies and software providers, and all VoIP application providers, argue that functional approaches or factor approaches are doomed to obsolescence as technology develops, and that the Commission should instead broadly classify services using IP technology, or at least those reaching or leaving the customer in IP format, as information services.

Another strain of comments advocates a layered approach to regulation. These commenters argue that the primary benefit of using IP to transmit voice is that it allows industry to move from using networks that are optimized for and dedicated to a single function, voice, to a network capable of delivering multiple functions. Therefore, regulation should reflect the fact that services and applications are no longer tied to the physical infrastructure. If dozens or hundreds of competing services and voice applications are provided over the infrastructure layer, there is little or no justification for continued common carrier regulation at those levels. Rather, they argue that the focus of common carrier regulation, if any, should be on underlying facilities, where issues of market power might still exist.

Interestingly, differences on classification among commenters did not necessarily translate to differences over jurisdiction. Some rural ILECs, their trade organizations, many of the commenting state commissions and NARUC argue that VoIP applications, if they are classified as telecommunications services, can and should be regulated at the state level. Other rural ILECs, the Federation for Economically Rational Utility Policy, and virtually all companies interested in offering VoIP applications, whether ILEC, IXC, CLEC, VoIP provider or other high tech company, have argued that VoIP applications are inherently interstate—that it is impossible to determine geographic end points for calls when customers can use VoIP applications from anywhere in the world, that IP networks ignore domestic and international boundaries when transporting bits, thus rendering the intrastate/interstate distinction meaningless, and that the Internet and services provided over it have always been considered to be subject to federal jurisdiction only.

With regard to whether economic common carrier regulation should apply, high tech companies and VoIP application providers overwhelmingly also agreed that there is no need for it. Many commenters that argued some VoIP applications should be classified as telecommunication services, nevertheless, also argued that they should be subject to federal jurisdiction only and that the Commission should forbear from applying economic common carrier regulation. The Illinois Commission, while arguing that state and federal regulation should coexist, with preemption only applying to state requirements that are inconsistent with federal requirements, nevertheless thought that extension of traditional utility regulation to emerging IP-enabled services was unwarranted. Some state commissions and many commenting rural ILECs concluded that VoIP applications should be subject to the same level of regulation as traditional voice providers, although America's Rural Consortium pointed out that this parity could be achieved through federal preemption of state regulation of voice service and removal of regulations from both VoIP and traditional providers.

There was general agreement among the commenters that universal service, 911 and other social obligations of this type will continue to be important in the new environment. There was, however, disagreement as to how best to achieve these goals. VoIP application providers and many of the technology-oriented trade groups tended to argue that obligations like access to 911 should only be made mandatory over time in response to a market failure, and that there has already been significant progress through voluntary industry action. They also argued that universal service and access charges should not apply until broader reforms to these systems are completed, as otherwise the Commission would impose unsustainable systems on a new technology. Others argue for mandatory application of these requirements, with most commenters focusing on specific areas: groups involved with advocating for disabilities access argue that mandatory disabilities access requirements should apply; some incumbent and rural ILECs that receive support from the Universal Service Fund and access charges argue that these obligations should apply pending changes in the system.

The Commission has received a wealth of comments that truly represent views across the spectrum.

IV. RECENT ACTIONS

In addition to our work on the IP-Enabled Services Proceeding, the Commission also is working on several petitions regarding VoIP in addition to recently resolved petitions.

The Commission recently resolved the following petitions:

- *Pulver.com.* As previously mentioned, on February 19, 2004, the Commission released an order resolving a petition for declaratory ruling filed by Pulver.com. In that order, the Commission found that Pulver.com's Free World Dialup Service was neither telecommunications nor a telecommunications service, but was instead an information service subject to federal jurisdiction, and that state regulation conflicting with this classification would most likely be preempted. This order was significant in terms of clearly establishing that Internet-only voice applications would be treated very much like any other applications traveling over the Internet: as being unfettered by federal or state regulation.

- *AT&T.* On April 21, 2004, the Commission released an order resolving a petition for declaratory ruling filed by AT&T. In this order, the Commission denied AT&T's request to exempt from access charges its use of VoIP in providing voice service where AT&T only used the technology to transport calls that originated and terminated on the public switched telephone network, and did not provide any enhanced functionality, cost savings, or net protocol conversion for the end user. This transport was carried out as part of AT&T's conventional service offerings and was transparent to the consumer. The Commission, by issuing this decision, did not prejudge the application of access charges to other types of VoIP service, which are still subject to consideration in both the IP-Enabled Services Proceeding and the Intercarrier Compensation docket. Thus, this decision was explicitly limited to the factual circumstances described by AT&T.

Petitions pending before the Commission are as follows:

- *Vonage.* On September 22, 2003, after the Minnesota Public Service Commission ruled that Vonage's service was a regulated telephone service under state law, Vonage filed a petition for preemption of this decision. Subsequently, Vonage obtained a reversal of this decision from a federal district court. An appeal of that court decision to the United States Court of Appeals for the Eighth Circuit is pending, while Vonage's preemption petition is still pending before the Commission.
- *Level 3.* On December 23, 2003, Level 3 filed a petition for forbearance, requesting that the Commission forbear from applying access charges to calls that originate or terminate as Internet protocol calls on one end, with the other end originating or terminating over the public switched telephone network. Level 3 excluded from its petition those areas served by rural ILECs as defined in section 251(f)(1) of the Communications Act. The twelve month deadline for Commission action in this proceeding is December 23, 2004, with a possible extension of three months beyond that date.
- *SBC.* On February 5, 2004, SBC filed a petition for forbearance asking the Commission to find that services and applications provided over Internet protocol platforms are information services subject only to federal jurisdiction, and as such to forbear entirely from applying Title II common carrier regulation to such services. The twelve month deadline for Commission action in this proceeding is February 5, 2005, with a possible extension of three months beyond that date.
- *Inflexion.* On February 27, 2004, Inflexion filed a petition for declaratory ruling, asking the Commission to find that calls made to or from Inflexion's VoIP service in areas that it characterizes as underserved are exempt from access charges. Inflexion's definition of underserved areas incorporates areas served by rural ILECs that Level 3 explicitly declined to cover in its petition.

In addition to the IP Enabled NPRM, these petitions also present opportunities to resolve specific questions related to VoIP. In addition, many of the issues that relate to universal service and intercarrier compensation are being considered by the Commission in other proceedings. Moreover, the Commission expects to release in the near term a Notice of Proposed Rulemaking addressing issues regarding VoIP and the Communications Assistance for Law Enforcement Act ("CALEA") raised by the Department of Justice, the Federal Bureau of Investigation, and the Drug Enforcement Agency in their recently filed petition for rulemaking. Consideration of VoIP issues will not delay broader resolution of those dockets, and the Commission hopes to move expeditiously on all fronts.

V. CONCLUSION

The Commission is very aware that VoIP is leading to significant developments in telecommunications markets challenging traditional industry economics as well as traditional regulatory institutions and processes. Perhaps most importantly, from the perspective of a regulator, VoIP is changing the nature of the relationship between consumers and providers. It would be irresponsible, as well as counterproductive, for any regulator to impose obsolete regulations reflexively, simply in order to protect a legacy regime. The examples of mobile wireless service and the Internet are perhaps most instructive in this respect. In both cases, the technologies have developed free of many of the regulatory requirements and regimes applicable to traditional monopoly common carriers, notwithstanding long and hard fought battles to impose such requirements. Indeed, it took an Act of Congress before the FCC could preempt counterproductive state regulation of cellular service. Today, the

American consumer and economy are far better off because of the deregulatory course that helped these two industries develop, innovate, expand and now touch millions of lives, brining considerable benefits to consumers, and generating substantial economic growth. All without traditional common carrier utility regulation.

The Commission's decisions regarding VoIP will have the farthest-reaching consequences of anything the Commission will consider in the near future. The Commission is considering nothing less than the future of electronic communications for today's and future generations. Consumers, the many industries that rely on information technology and advanced communications in their business, the telecommunications, computer and software industries, and the investment community are all counting on the Commission to get it right. It also is not an overstatement to say that the world, also, is watching how the U.S. decides to treat these services. Telecommunications regulators and policy makers in other countries want to know whether the United States will create an environment that is conducive to growth and investment in innovation, or an environment where the United States becomes mired in reflexive, legacy regulation and regulatory processes that stifle progress.

Clearly, I believe we should look forward rather than backwards. When dealing with revolutionary new technologies we need to start from the perspective of how to best create the world we all want to live in, rather than applying tired regulations quickly being rendered obsolete. The Commission, however, is constrained by the Act, which divides the world into regulated telecommunications services and unregulated information services. While the Commission certainly has some ability to fine tune treatment of new technologies given its discretion and the flexibility granted to it by Congress, the Commission is still constrained by this structure. If you believe that VoIP and other new technologies are transforming the telecommunications market in ways that cry out for new regulatory approaches, you may need to consider whether the tools the Commission has today are appropriate for the task.

In the meantime, the Commission will continue forward, and the guidance and leadership of Congress is crucial to the success of its process. On behalf of the FCC, I want to thank you, Mr. Chairman, for calling this hearing, and we look forward to working with you and other members on these issues.

1 The Commission did receive a petition regarding VoIP services as early as 1996, and received another following the release of its 1998 report to Congress regarding universal service, often called the "Stevens Report." There was not, however, any consequential activity following these petitions.

Mr. CANNON. Thank you, Mr. Pepper.

Mr. Langhauser.

TESTIMONY OF JOHN LANGHAUSER, ESQ., VICE PRESIDENT, LAW, AND CHIEF COUNSEL, CONSUMER SERVICES GROUP, AT&T CORPORATION

Mr. LANGHAUSER. Mr. Chairman, Congressman Chabot, thank you very much for giving me the opportunity today to discuss voice-over-Internet protocol.

AT&T intends to provide IP-based services to all of our key markets. In March of this year, we launched our residential VoIP service, known as AT&T CallVantage. Today, it is offered in 32 States and Washington, D.C. That is 100 major markets in 4 months. Voice-over-IP is a foundation for our future. Indeed, because of recent Federal policy changes concerning unbundled network elements, VoIP will soon become AT&T's only viable alternative for offering new competitive local service, but unfortunately, only for those customers who can obtain and afford broadband.

Much of Silicon Valley will benefit from an IP explosion. Small businesses will profit from a portable VoIP services. The resulting productivity gains can, in turn, drive broader economic growth. These benefits will only emerge if policymakers bring certainty and stability to the regulatory rules surrounding VoIP. It should be regulated with a light hand at the Federal level. In particular, it should not be saddled with the current, flawed intercarrier compensation markets.

VoIP cannot be allowed to develop into yet another Bell-controlled technology. AT&T's ability to compete for customers and invest in VoIP will be hampered if the Bells are allowed to continue such anticompetitive practices as refusing to sell broadband to customers purchasing voice services from a competitor.

Let me provide more details: VoIP holds the promise of choices and capabilities far beyond today's circuit-switched offerings. In the IP environment, voice services and futures can be provided and enhanced much more efficiently. VoIP could well become the killer application that drives broadband adoption.

AT&T fully intends to lead the VoIP revolution. We have invested heavily to upgrade our total network, including some $3 billion last year alone. Our consumer offer includes advanced features such as the ability to check voice mail from your computer and dynamically control your feature settings yourself.

AT&T has long been committed to providing a choice for local telephone service. Today, we provide local service to about 4.7 million residential customers and 4.5 million business lines. Virtually all the residential customers are served using Uni-P. But VoIP, which requires broadband, is not an option for the majority of our current local customers.

Legislative and regulatory certainty, which fosters VoIP as an emerging technology, will encourage AT&T to invest in VoIP and remain in the domestic residential voice market. Congressman Pickering's bill provides for Federal regulation and access and universal service reform. Chairman Sensenbrenner and Congressman Conyers have offered legislation ensuring that the Telecommunications Act is not construed to supersede the antitrust laws.

We commend these efforts to restore the potential for a competitive communications marketplace. We agree with those who have said that VoIP must provide access for the disabled, 911 and must cooperate with requests from law enforcement. In contrast, the universal service and intercarrier compensation schemes of today are badly broken and require substantial revisions before they can or should be applied to VoIP.

The FCC's delay in reforming these regimes benefits the incumbents. Nothing about VoIP threatens universal service. The real threat is the shrinking base of interstate revenues that support the system today. AT&T has proposed moving to a flat rate charge for each telephone number, which would include VoIP, be competitively neutral and provide a solid foundation for the fund. The FCC has full authority to implement such reforms, but AT&T's petition has been pending for over 15 months.

Current access charge regulations are especially unworkable, but the FCC's long-promised overhaul of intercarrier compensation has yet to occur. VoIP collectively serves several hundred thousand customers nationwide, while the Bells serve nearly 100 million. It makes no sense to require nascent VoIP providers to subsidize the monopoly local carriers. Nobody demanded that the auto industry subsidize the buggy manufacturers or the computer industry the typewriter providers.

If VoIP is to deliver on its promising potential, then, it cannot be regulated like plain old telephone service. Today, we are asking for your support to keep that from happening so that all Americans can realize the competitive and innovation benefits of VoIP technology.

Thank you again for inviting me here today, and I look forward to your questions.

[The prepared statement of Mr. Langhauser follows:]

PREPARED STATEMENT OF JOHN J. LANGHAUSER

Mr. Chairman and Members of the Committee, thank you very much for giving me the opportunity today to discuss Voice Over Internet Protocol. AT&T intends to provide IP-based services to all of the key market segments—large enterprises, call centers, small offices, teleworkers, and residential users. We've been delivering Business IP services since 1997, and in March 2004, AT&T launched its residential VoIP service, known as AT&T CallVantage^{sm} Service. Today it is offered in 32 states and Washington D.C.—that's 100 major markets in less than four months.

VoIP is the convergence of voice and data, with the potential to bring choice and innovation to the telecommunications marketplace. If allowed to grow unimpeded by legacy regulation, it will offer consumers an increasing array of advanced features not available today to enhance ways of communicating and simplify busy lives.

VoIP will also contribute significantly to the business world. Teleworkers using VoIP will be far more productive and successful at their work. VoIP will bring the kind of advanced voice and data service now available only to Fortune 500 companies within the reach of small and medium-sized businesses. Much of Silicon Valley is now in the IP value chain and will benefit from an IP explosion in this market. The resulting productivity gains can, in turn, drive broader economic growth and raise standards of living for all Americans.

These benefits will only emerge, however, if policymakers act promptly to limit regulation to a light-handed regime that allows VoIP to develop free of burdensome regulation at the federal, state or local level. Imposing today's inflated access charges on nascent VoIP providers would severely impede the growth of VoIP. VoIP providers are already paying substantial compensation to local exchange carriers for the right to terminate traffic on their networks. They should not have to subsidize their established competitors as well. With respect to intercarrier compensation, the priority should be on reform rather than burdening innovative new services and technologies with an outmoded regulatory model heavy with subsidies.

VoIP seeks only the favorable regulatory treatment that other emerging voice technologies have received. Relieving wireless carriers of much incumbent economic regulation led to amazing increases in investment, innovation, and consumer adoption. While the FCC authorized commercial cellular services in 1981, in 1992 there were only nine million subscribers. It was only when Congress empowered the FCC in 1993 to forbear from imposing legacy regulation on cellular providers and made significant additional spectrum available for their use, and the FCC exempted them from tariffing and entry and exit regulation, that wireless use exploded. By the end of 2002, there were 141.8 million subscribers nationwide.

Many questions regarding whether to foster VoIP's emergence as a competing technology or saddle it with legacy wireline regulation and stifle its development are currently before the FCC. Unless and until Congress acts, we believe it is incumbent on the FCC—indeed, consistent with its congressional mandate—to take steps to establish an appropriate regulatory framework that encourages investment and innovation. The FCC's unreasonable delays to date in resolving even the most preliminary regulatory issues surrounding VoIP do not meet the basic requirements of sound administrative procedure.

Firm resolve in enforcing the pro-competitive policies of the 1996 Act is a necessary first step on the path to VoIP. Business cases based on a "build it and they will come" approach to deploying mass-market local facilities have been almost uniform failures. Congress recognized this when it passed the 1996 Telecommunications Act and provided for resale and the unbundled network elements platform (UNE-P) to enable carriers to develop local subscriber bases which would support a migration to building their own local facilities. In both the business and residential markets, however, facilities-based service requires a significant concentration of demand to be economic. To the extent multiple networks can ever economically compete, a significant customer base is needed to justify network deployment and reduce the risk of such deployment. Today, AT&T provides local service to more than

4.3 million residential lines and 4.5 million business lines, including 1 million small business lines. We have done so through a combination of facilities-based entry—we have invested billions of dollars in our own local facilities since 1996—and the lease of Bell network elements.

In the wake of the regulatory certainty generated by the U.S. Supreme Court TELRIC decision and the highly contested FCC Triennial Review announcement in February 2003, AT&T entered local service in thirty-seven additional states for a total of forty-six states. However, in view of the regulatory uncertainty generated by this same Administration and FCC's decision not to appeal the D.C. Circuit reversal of the February 2003 order, AT&T has had to re-assess the business case for local and long distance residential markets. The re-introduction of regulatory uncertainty has strangled mass-market local competition in its very infancy.

AT&T strongly believed that the D.C. Circuit decision is both wrong and flatly contradicts Supreme Court precedent, but the Administration refused to appeal it. The Bell companies' refusal to negotiate reasonable interconnection and leasing agreements in the wake of that decision has left AT&T no choice but to stop incurring the costs to solicit new local phone customers in its residential markets. With the Bell companies poised to raise wholesale rates for UNE-P as early as November, we will simply not be able to provide a bundle of local and long distance services economically and build the customer base that so greatly facilitates our VoIP deployment.

Without appropriate legislative and regulatory treatment, VoIP could develop into yet another technology controlled by the Bells. Without competition, the Bells may digitize voice but have no incentive to develop the myriad software applications for advanced and converging features that truly promise to change the way we communicate. Remember that these are the same companies that held back the deployment of DSL services to residential customers for some ten years so customers would have to take their other, higher priced services. Only when forced by competition, in that case the deployment of broadband Internet connections by cable operators and competitive carriers Covad and Rhythms, did the Bells finally introduce mass-market, high-speed Internet access service. Similarly, without the threat of losing customers to a VoIP rival, the Bells will have no incentive to invest in and deploy this new technology or the rich array of features it is capable of providing.

The prospects for competition will be thwarted, if the Bells are allowed to continue such anticompetitive practices as refusing to sell their broadband service to customers that purchase voice service from a competitor, or requiring their broadband customer to purchase a local exchange line as well. The Bells' ability to restrict broadband customers from subscribing to anyone else's voice services has attracted widespread attention and many states have sought to prohibit these anticompetitive practices—but they continue. Unless we and other competitors are allowed—quickly—to fairly compete for voice customers, we will not be able to invest in VoIP, and VoIP will become just another Bell-controlled technology.

Legislation proposed by Chairman Sensenbrenner and Congressmen Conyers would greatly further the goal of competition and protect against the incumbents' anticompetitive practices by reaffirming the application of the antitrust laws to the telecommunications sector. It would prevent the Bells from attempting to perpetuate their monopolies by unlawful tying or refusing to share network facilities with competitors at reasonable prices. AT&T strongly endorses this legislation.

Let me provide more detail on each of these points.

VOIP HOLDS THE PROMISE OF NEW CHOICES AND MORE CAPABILITIES

VoIP holds the promise of choices and capabilities far beyond today's circuit-switched offerings. It enables consumers to enhance and tailor their communications services to their needs and lifestyles at competitive prices. It very well could be the "killer app" to drive widespread broadband adoption for which we have all waited. It could also be an important economic driver for our nation.

AT&T fully intends to lead the VoIP revolution for businesses and consumers. We have invested heavily to upgrade our total network, including some $3 billion in 2003 alone, and we have already met our goal of providing VoIP service in the top 100 markets in the country this year.

With VoIP, voice service is just another "hosted application" like e-mail, letting customers take their phone numbers wherever they go and access connections over any device, such as a standard home telephone, wireless phone, or computer. AT&T's consumer offer, AT&T CallVantage[sm] Service, for example, already includes a host of new advanced features and the ability for consumers to dynamically tailor and control their feature settings via website or telephone any time day or night as often as they want. Advanced features include advanced call forwarding features and "do not disturb" options that enable consumers to program the service so that the phone answers to their needs instead of the other way around. AT&T CallVantage^{sm} Service provides subscribers a "Personal Call Manager Web Site," which gives subscribers complete, dynamic control over their answering, voice mail and other capabilities. Subscribers can check their voicemail from their computer and forward information as a "talking" e-mail. Innovations, and the resulting benefits to consumers, will only increase as device manufacturers, network operators, service providers and application developers take full advantage of the ability to integrate voice, data and advanced computer capabilities.

In the IP environment, voice services can also be provided much more efficiently. IP technology allows for more efficient routing of calls than traditional circuit-switching. These efficiencies enable more innovative service packages. Current VoIP offerings allow customers that have a broadband connection to place unlimited calls anywhere in the country for a single, low monthly price. The Alexis de Tocqueville Institution concluded earlier this year that government at all levels could save $3–10 billion annually—up to 60% of their current phone bills—by replacing circuit-switched service with VoIP. You should not, however, think of VoIP as "cheap phone service." It promises to be lower-cost, yes, but with a host of new communications management features and options that go well beyond today's "plain old telephone service" ("POTS").

A "HANDS-OFF" APPROACH IS THE APPROPRIATE REGULATORY APPROACH FOR VOIP

Allowing VoIP to develop in the marketplace is a critical step to bringing this Nation into the digital age. AT&T welcomes the fact that many Members of Congress support a "hands off" approach to VoIP and have introduced legislation that would bring the benefits of competition and innovation to the telecommunications marketplace. Congressman Pickering, for example, has proposed a deregulatory approach to VoIP that acknowledges the need to reform the current subsidy system and allow this nascent service to flourish.

Fundamentally, VoIP legislation must recognize that because the Internet is global in nature and these services will be deployed nationwide, a federal framework makes the most sense. Forcing U.S. VoIP providers to develop 50 different varieties of VoIP services to comply with a patchwork of potentially inconsistent state regulatory burdens could hinder their development. Continuing regulatory uncertainty as to federal versus state regulation of VoIP, or worse yet, the regulatory uncertainty that would accompany implementation of 50 different regimes to regulate VoIP, would inevitably impede investment, in direct opposition to the federal policy of creating a regulatory framework that promotes the growth and development of broadband services. Indeed, recognizing the critical importance of a uniform, nationwide deregulatory environment, the Pickering bill prohibits even the FCC from regulating VoIP applications except as specifically authorized.

Such an approach will be critical to VoIP's ability to lead the United States' broadband revolution: the United States' broadband penetration lags behind that of a number of other countries. Many of those who have higher rates of broadband penetration have recognized that allowing VoIP to flourish will contribute to a positive economy and allow them a competitive edge in the global marketplace. The United States, too, must protect its economic interests by abandoning outdated policies favoring and protecting incumbent revenue streams.

Allowing emerging VoIP services to develop free of unwarranted, legacy regulation allows carriers to design the service to respond to customer needs and interests, and to remain flexible in their business plans as customer preferences emerge, rather than be bound by a government-dictated vision of what the service should include and what is a benefit to consumers. As FCC Chairman Powell stated on February 8, 2004:

> the case for government imposed regulations regarding the use or provision of broadband content, applications and devices is unconvincing and speculative. Government regulation of the terms and conditions of private contracts is the most fundamental intrusion on free markets and potentially destructive, particularly where innovation and experimentation are hallmarks of an emerging market.

The wisdom of this approach was confirmed recently—in reverse—when a new local VoIP provider concluded it could not stay in business in any of the states in which it had been operating when faced with an order from Washington state regulators to register as a telephone company and comply with the same laws as other long distance companies (including the payment of access charges). Regulators must be able to approach VoIP service flexibly if they expect VoIP to bring its promised benefits to consumers and competition.

We agree with those who've said that providers of VoIP services must meet important social policies. Providing access for the disabled, enabling public safety (911) response, and cooperating with lawful requests for information from law enforcement are issues that the industry can and is working to resolve, and AT&T is taking a lead in these efforts. While government has a legitimate role in ensuring that these things get done, it should refrain from regulating this new service in these or other areas in the absence of a demonstrated failure on the part of industry to act appropriately. We may also need some flexibility and reasonable transition periods to achieve these policy goals, in recognition of the fact that IP-enabled services present different technical and operational issues than those considered when the legacy common carrier regulations were originally developed. Nonetheless, we believe that the enormous flexibility and power of VoIP promises to address these issues in ways superior to current circuit-switched technology.

Other legacy regulations, however, will require substantial revisions before they can or should be applied to VoIP. The universal service and intercarrier compensation schemes are irremediably broken and indeed, no longer make sense even in the context of the traditional, circuit-switched wireline telephone services for which they were developed. Prompt attention to these fundamental flaws in existing regulation is urgently needed so that IP-enabled services are not burdened with costly and outdated, broken regulatory schemes that would prevent VoIP services from reaching their potential.

Let me emphasize that nothing about VoIP threatens universal service. The problem with the universal service fund (USF) is that it is still supported by a shrinking base of interstate revenues for traditional telecommunications services. A growing fund with a shrinking base cannot be sustained. It's long past time for the universal service systems in this country to be reformed, and we support VoIP being part of the broader reform of the USF system. We think VoIP providers should contribute to a reformed universal service system—in a sustainable, fair, and nondiscriminatory manner.

AT&T has proposed a contribution system to the FCC that would replace the current revenues-based system with a numbers/capacity-based system that is fairer and more sustainable. Under our proposal, providers would pay a flat-rated charge for each assigned telephone number that maps to a unique end-user's service. Services known as "special access services" would also be assessed a flat-rated charge based on the capacity of the service. Such a system would be competitively neutral, and would provide a solid foundation for the fund because the use of numbers is increasing. Moreover, VoIP providers would be fully included, since their service nearly always uses traditional phone numbers—as would future technologies, which are likely to retain the use of numbering. The Commission has full authority to implement such reforms—but it has yet to do so. In fact, it has delayed action on every major VoIP issue it has confronted thus far. It took the FCC 18 months to decide the merits of a petition AT&T filed—and nearly as long to rule on a similar one filed by pulver.com—regarding the regulatory consequences of offering VoIP services. Such delay fails to meet basic notions of fairness in administrative procedure—and harms competition. Carriers need clarity and predictability in the marketplace if they are to make the risky investment needed to make VoIP widely available.

Especially unworkable and in need of attention are the Commission's vastly outdated access charge regulations. The access charge scheme was developed decades ago to ensure that whenever a long distance company used the local network, it would subsidize local service by paying grossly inflated rates to the local carrier. While there was much in this framework to which one could object, it remained workable as long as local carriers and long distance carriers operated in separate markets. Its infirmities became apparent and unsustainable when those carriers entered each others' markets, and even more so when wireless companies and ISPs became the largest users of access minutes. For that reason, eight years ago, Congress ordered that implicit subsidies, including those in access charges, must be eliminated. Unfortunately, they still remain in place eight years later, and the FCC's long-promised overhaul of its intercarrier compensation regime has yet to occur. While Chairman Powell commendably opened a proceeding examining needed revisions as one of his first acts as Chairman, that docket remains unresolved more than three years later.

Now, the emergence of VoIP services dramatically underscores the urgent need for the Commission to meet its responsibilities under the APA and complete intercarrier compensation reform. Whatever the historical wisdom of requiring interexchange carriers to subsidize through inflated access charges local exchange carriers operating in a different market, it makes no sense to require nascent VoIP providers to subsidize the monopoly local exchange carriers against whom they will be directly competing. VoIP providers collectively serve only several hundred thousand customers, while the Bells serve nearly one hundred million. Having VoIP providers subsidize the incumbents cannot be the right answer. No one demanded that the auto industry subsidize the buggy manufacturers, or the computer industry the typewriter providers, or email the post office.

The far better course is comprehensive reform of the intercarrier compensation regime to eliminate market distortions and opportunities for regulatory arbitrage. Nearly every segment of industry agrees that there is a need to move to a rational system in which all traffic is exchanged under the same compensation rules. Even OPASTCO—the Organization for the Promotion and Advancement of Small Telecommunications Companies—acknowledges the need for intercarrier compensation reform, although its members directly benefit from current law. In a hearing before the Senate Commerce Committee on June 16, 2004, Arturo Macias, current Chairman of OPASTCO, testified that although it was important for rural carriers to be able to recover their costs of providing access to their networks, current intercarrier compensation rates are not cost-based, and OPASTCO would not oppose their reform.

Until that reform occurs, however, these legacy access charges should not apply to IP-enabled services, even on an interim basis. Even Qwest agrees with us that providers using IP at either the origination or termination points of telephone traffic should not pay access charges, even if the traffic at some point traverses the public switched telephone network. The imposition of above-cost access charges on IP telephony would radically alter the economics of providing VoIP services and would severely impede the development of those services.

Contrary to the Bells' claims, VoIP providers do not get a "free ride" when they don't pay access charges. To the contrary, VoIP providers typically purchase what are known as Primary Rate Interfaces ("PRIs")—a type of high-speed line—or other local business lines to connect to the public switched telephone network, and they pay for termination as an enhanced service.

AT&T agrees that affordable service needs to be maintained in high-cost areas of the country. Applying the legacy access charge regime to VoIP, however, is not the way to achieve this result and would prove counterproductive and market-distorting. It simply slows the deployment of new and desirable technologies while driving users away.

Today we are at a crossroads where we must call upon your leadership. If VoIP is to deliver on its promising potential—and offer something truly different in the marketplace—then it cannot be treated and regulated like plain old telephone service. We are asking for your support to keep that from happening, so that Americans can finally realize the long-promised benefits of widespread competition and the innovations promised by VoIP.

Thank you again for inviting me here today, and I look forward to your questions.

Mr. CANNON. Thank you, Mr. Langhauser.

Mr. Cordi.

TESTIMONY OF STEPHEN M. CORDI, ESQ., CPA, DEPUTY COMPTROLLER FOR THE MARYLAND COMPTROLLER OF THE TREASURY, STATE OF MARYLAND

Mr. CORDI. Thank you, Mr. Chairman, Congressman Chabot. Good morning and thank you for the opportunity to testify on the regulatory aspects of voice-over-Internet protocol.

I am here on behalf of the Federation of Tax Administrators. The FTA is an association of tax agencies of all 50 States, New York City and the District of Columbia. My comments today will be limited to the State and local aspects and preemptions found in H.R. 4129, a bill that has been referred to this Committee, and I will leave the regulatory matters to those with expertise in those areas.

We have four major objections to the preemption of State tax authority found in H.R. 4129: it discriminates against other providers of voice communications services; It represents a considerable fiscal cost to the State governments; it runs completely counter to the country's established system of federalism; and no case has been made for preempting State and local tax authority.

Our reasoning for this is as follows: first, voice-over-IP is an exciting new technology, and it is always tempting to want to nurture a new product. But in doing so, we must not forget existing and competing products. One of the primary goals of tax policy is to treat similar taxpayers and similar goods and services in a similar fashion. Government should not choose the winners and losers in the marketplace through tax policy.

One thing is clear: preempting State taxation of voice-over-IP services will put land phone services and wireless phone services at a competitive disadvantage. The technologies are different, but they are functional equivalents. All three industries provide voice communications services. All three, and those that will emerge in the future, should be taxed in a similar manner. Preempting State taxing authority with respect to voice-over-IP goes 180 degrees in the wrong direction.

Secondly, State and local governments currently collect about $10 billion annually on sales of telecommunications services. The Congressional Budget Office has estimated that preempting the taxation of voice-over-IP could reduce State revenues by at least $3 billion a year within 5 years, and that may be, as the Chairman said in his opening remarks, an underestimation of the growth of voice-over-IP. And we anticipate that preemption would also accelerate the growth of voice-over-IP and quickly lead to the loss of much of the remainder of the $10 billion.

Beyond that, it is possible that H.R. 4129, as written, would also prohibit the States from collecting some substantial part of the $7 billion we now collect in property, income and sales taxes from existing telecommunications providers as assets are shifted to voice-over-IP. In short, preempting the taxation of voice-over-IP services will have a major and adverse impact on State and local fiscal systems and constitute a de facto repeal by the Congress of a source of taxation available to State and local governments for over a century.

Third, broad preemption of State tax authority to tax voice-over-IP services will represent a radical departure from historical practice for Congress. Both the States and Federal Government are sovereign entities with the right to tax. Congress has heretofore generally limited preemption of State and local taxation to narrow situations where there has been an excessive reporting burden or a compelling need for uniformity.

Finally, not only is this a uniquely broad preemption, but no evidence suggests that there is a compelling national interest in eliminating the State taxation of this technology. It has certainly not been showing of a need for preemption on the basis of complexity or lack of uniformity. There may indeed be bona fide issues that need to be resolved on how State and local taxes should be applied to voice-over-IP services. Any new type of business creates the need for new regulations and policy adjustments. But it certainly seems excessive to preempt the better part of an entire tax on the theory that there may be issues that need to be resolved.

Any issues can best be dealt with through an honest and constructive dialogue involving all affected parties. And in conclusion, voice-over-IP services hold significant potential to improve our society. Congress can promote competition, preserve State tax authority and protect the public interest by refraining from any policy that unnecessarily preempts State and local taxing authority, discriminates against traditional voice communication providers and disrupts State and local fiscal systems.

Thank you.

[The prepared statement of Mr. Cordi follows:]

PREPARED STATEMENT OF STEPHEN M. CORDI

Mr. Chairman and Members of the Committee:

Thank you for the opportunity to appear before you today on the important question of the appropriate federal policy regarding the regulation and taxation of Voice over Internet Protocol (VoIP) technology. My name is Stephen M. Cordi. I am the Deputy Comptroller for the State of Maryland, and I appear before you today on behalf of the Federation of Tax Administrators, an association of the principal state tax administration officials from the 50 states, D.C. and New York City.[1] I am the Immediate Past President of the Federation.

My comments today will be limited primarily to the issue of potential federal legislation that would eliminate, limit or otherwise preempt the ability of state and local governments to impose taxes on VoIP services. There are important issues involving potential federal preemption of state authority to regulate VoIP services, but I leave those to others with expertise in the area. Further, I will direct my comments principally to the state and local taxation provisions in H.R. 4129, The VoIP Regulatory Reform Act of 2004, that was introduced by Rep. Pickering and others since that is the clearest expression of potential federal policy in existence today.[2]

The thrust of my comments today can be summarized as follows: Congress should not take action at this time that would preempt the ability of state and local governments to impose taxes on VoIP communications services. Such an action would discriminate against other providers of voice communications services using technologies that are subject to tax and would deprive states and localities of significant amounts of revenue in the very near future. In addition, such an action would run counter to our system of federalism and to the traditional Congressional posture of not intervening in state taxing matters. Finally, we believe that no case has been made that would warrant federal intervention at this point, and that federal action of the sort envisioned in H.R. 4129 would obviate any possibility of a cooperative state-industry dialogue to identify and resolve any issues that may be present in state and local taxation of VoIP services.

FEDERAL PREEMPTION WOULD CREATE DISCRIMINATORY SYSTEM

There is no doubt that VoIP is an exciting new technology that holds significant potential to provide enhanced, more convenient communications services to some consumers and businesses at costs that are sometimes lower than they face today. Each week seems to bring the announcement of another VoIP offering, not only from start-up companies, but also from established telecommunications companies of all types.[3] At its core, however, we must remember that VoIP is one of several competing technologies that can be used for providing voice communications services.

One of the primary goals of tax policy is to treat similar taxpayers and similar goods or services in a similar fashion when it comes to taxation. Only by taxing similar or functionally equivalent services in the same fashion, can we ensure that consumer choices are based on price and quality of service and not distorted by tax policy. Preempting state and local taxation of VoIP services as proposed in H.R.

[1] The Federation of Tax Administrators is an association of the state tax agencies in the 50 states, District of Columbia and New York City with principal programs in information exchange, training and intergovernmental coordination. FTA policy regarding federal preemption of state taxing authority was adopted by the membership at its 2004 Annual Meeting. That policy statement is attached.

[2] H.R. 4129 would, among other things, prevent any state or political subdivision from imposing any tax, fee or other charge on the offering or provision of VoIP services. It would also preempt any state regulation of VoIP services and would limit the extent to which the Federal Communications Commission could regulate VoIP services.

[3] There are several types of VoIP services and a variety of consumer features available from various VoIP providers. Some VoIP services do not use the publicly switched telephone network (PSTN), but estimates are that currently 90 percent of all VoIP calls either originate or terminate on the PSTN.

4129 would create an unprecedented tax preference for one form of voice communications services (VoIP), and it would place other traditional land-line and wireless voice providers at a substantial competitive disadvantage because they would still be obligated for existing state and local taxes. Such a policy creates an unlevel playing field that works against those providers not employing VoIP and will cause a misallocation of resources in the economy. Enacting such a discriminatory arrangement will undoubtedly create additional calls for federal intervention in an effort "to level the playing field."

In considering the appropriate tax policy for VoIP, Congress must consider function over form. That is, the function of VoIP is to provide voice communications services, and it is the functional equivalent of other forms of voice communications services. It should be taxed in a manner similar to other voice communications services to avoid distorting consumer choices and to avoid placing Congress in the position of choosing winners and losers from among competing telecommunications providers. H.R. 4129 runs directly counter to that proposition.

If Congress chooses to base its tax policy decisions on the technology employed in VoIP services, rather than the function of VoIP, it is likely to find itself continually one step behind the technology curve and facing a continuing set of requests for intervention. A prime example of this result is the passage of the Internet Tax Freedom Act in 1998 that was written when dial-up access was the predominant, if not exclusive, method of providing Internet access. Within a relatively short period of time, however, other technologies developed and not all were treated in the same manner under the federal law as juxtaposed against state tax systems. This led to demands for further interventions and preemptions by the Congress as it considered extending the Act this year.

In short, preempting state and local taxation of VoIP services, while leaving the taxation of other forms of voice communication intact, constitutes an unsound tax policy that discriminates against traditional voice communication providers. This is not to suggest that there are not likely bona fide issues of the manner in which state and local taxes should be applied to VoIP services. Such issues can only be identified and resolved through an honest and constructive dialogue among the affected parties. Adoption of policies such as those contained in H.R. 4129 would prevent such a dialogue from occurring and create a discriminatory tax environment.

FEDERAL PREEMPTION WOULD HAVE A SUBSTANTIAL REVENUE IMPACT ON STATES AND LOCALITIES

According to the Congressional Budget Office, state and local governments collect about $10 billion annually in general purpose transaction taxes (including sales taxes and telecommunications excise taxes) on sales of telecommunications services at the present time.[4] Further, CBO estimates that under *current* projections, it is expected that up to one-third of traditional voice traffic would migrate to VoIP within five years, thus implying a revenue loss to states and localities of upwards of $3 billion annually by that time. Enacting a tax exemption for VoIP services would undoubtedly accelerate that revenue loss and lead to the loss of a substantial portion of the $10 billion in a relatively short period of time.

In addition, depending on interpretations of the breadth of the tax preemption in H.R. 4129 as well as the interpretation of the state prohibition on regulating VoIP services in the bill,[5] a substantial portion of the $7 billion that CBO estimates states and localities collect from business taxes (property taxes, business profits taxes, and taxes on purchases) on telecommunications providers could be preempted as well.[6] That is, as assets of traditional telecommunications providers are shifted to VoIP services or are taken out of service due to the migration of traffic to VoIP providers, revenue from these business taxes will also be lost to state and local governments.

In short, a broad preemption of state and local taxation of VoIP services would have a substantial detrimental revenue impact on states and political subdivisions. It would, in fact, constitute a de facto repeal by the Congress of an entire category

[4] Letter to Senator Lamar Alexander from CBO Director Douglas Holz-Eakin regarding S. 150, the "Internet Tax Nondiscrimination Act," dated February 13, 2004. This does not include about $3–4 billion in 911 and Universal Service Fund fees that would be preempted under the bill as well.

[5] In the bill "regulate" is defined to mean "any governmental action that restricts, prohibits, limits or burdens, or imposes any obstacle, obligation or duty, or interferes with, [a VoIP] application

[6] For further discussion, see Michael Mazerov, "Proposed 'Voice over Internet Protocol Regulatory Freedom Act' Threatens to Strip States and Localities of billions of Dollars In Annual Tax Revenues, Center on Budget and Policy Priorities, Washington, D.C., July 20, 2004.

of taxes on which states and localities have long relied—taxes on telecommunications services and providers. States and localities would have two alternatives to deal with the preemption: reduce expenditures or raise the revenues from other taxpayers. Given that approximately 55 percent of all state and local expenditures are for education, social services and public safety, the impact of expenditure reductions will likely be felt in services considered critical by the citizens.[7]

FEDERAL PREEMPTION WOULD RUN COUNTER TO OUR SYSTEM OF FEDERALISM

Our system of federalism is founded on the concept that both the states and the federal government are sovereign entities and that both possess the sovereign ability to tax. The shared sovereignty with regard to taxation is a core element of political sovereignty. Moreover, our system is based on a precept that state and local elected officials, respecting the safeguards afforded all citizens by the U.S. Constitution, are in the best position to determine the appropriate tax policy for their citizens and for economic activity occurring within their borders.

Despite its plenary authority to regulate interstate commerce, Congress historically has been respectful of state tax sovereignty and has substantially limited the instances in which it has preempted state taxing authority.[8] Congressional preemptions (beyond those assuring respect for the Supremacy Clause) have generally been limited to relatively narrow areas where there has been a substantial showing of excessive burden or need for uniformity. Examples include the individual income tax treatment of workers in interstate commerce, treatment of nonresident pension income and property taxation of certain interstate transportation industries. In addition, Congress has in some instances fostered state tax sovereignty. Examples include the federal Tax Injunction Act that prohibits the federal courts from restraining the collection of a state tax where an adequate remedy exists in the state courts and the Mobile Telecommunications Sourcing Act that endorsed a resolution to the need for a single rule in sourcing wireless telecommunications services that was developed by the industry and the states.

Enactment of H.R. 4129 or similar policies preempting states from taxing a particular technology would represent a substantial departure from traditional Congressional positions and our federal system. Congress would be substituting its judgment for the judgment of state and local elected officials and effectively determining that states and localities should no longer tax voice communications services.[9] This stands in sharp contrast to the rich tradition of federalism on which our government was founded and which has served our country well. As our national and state economies have evolved, states have developed their tax policies with an eye toward accommodating new technologies as members of a stable marketplace. This system has worked well, and no evidence has been presented to suggest that state tax policies have impeded the growth of new technologies or state or national economies.

CASE FOR FEDERAL POLICY OF TAX PREEMPTION HAS NOT BEEN MADE

We believe that enacting the broad regulatory and tax preemptions contained in H.R. 4129 is unwarranted in that there has been no showing of a need for federal intervention.[10] Moreover, a policy of preemption would likely impede or preclude the development of sound long-term policy for VoIP that treats all voice telecommuni-

---

[7] U.S. Bureau of Census, Preliminary Estimate, State and Local Government Finance, 2002 Census of Governments, found at *http://www.census.gov/govs/www/estimate02.html.*

[8] For a more complete discussion (as well as an evaluation of certain current federal preemption proposals), see Charles E. McLure, Jr. and Walter Hellerstein, "Congressional Intervention to State Taxation: A Normative Analysis of Three Proposals," State Tax Notes, March 1, 2004.

[9] Most observers expect a rapid migration to VoIP even without a tax preference. Michael K. Powell, the chairman of the Federal Communications Commission, was quoted as saying, "We think pretty quickly there's no reason why virtually any communication service [won't be Internet-based]." Yuki Noguchi, "Identity Crisis," The Washington Post, Oct. 23, 2003. Preempting taxation of VoIP would constitute a de facto repeal of all taxes on voice telecommunications because all or nearly all forms of voice telecommunications would move to VoIP.

[10] The U.S. Senate has twice taken action to clarify that its actions are not intended to preempt state and local taxation of VoIP services. The Internet Tax Nondiscrimination Act (S. 150) as passed by the Senate in April 2004, contains a provision contained in a Manager's Amendment stating, "Nothing in the Act shall be construed to affect the imposition of taxon a charge for voice . . . service utilizing Internet protocol. . . ." On July 22, 2004, in a mark-up of its version of the "VoIP Regulatory Reform Act" (S. 2281), the Senate Commerce Committee approved an amended version of the bill that does not contain a preemption of state and local taxing authority and a dialogue with the sponsor of the bill established that the bill was not intended to preempt taxing authority.

cations providers in an equitable fashion and that is respectful of the tax sovereignty of the states.

The types of VoIP services that will be offered are still evolving as is the understanding of the issues involved in the taxation and regulation of VoIP. On the tax front, there has not, to my knowledge, been any attempt to demonstrate a need for federal preemption on the basis of complexity or lack of uniformity. A review of recent tax literature reveals only one article examining state tax issues associated with VoIP,[11] and the bulk of the issues identified in that piece involve whether VoIP would qualify as a telecommunications service under state tax statutes, not issues of complexity or uncertainty that would make a tax on VoIP services difficult to administer or comply with. While there may well be issues that should be addressed, we do not believe it is appropriate to preempt all state and local taxation on the theory that there may be issues to deal with. Through efforts such as the Mobile Telecommunications Sourcing Act and the Streamlined Sales Tax Project, states have shown their willingness and ability to work with stakeholders to address bona fide issues of complexity and uniformity. A broad federal preemption would preclude any such discussions.

CONCLUSION

VoIP services hold significant potential to provide consumers with more choices for voice communications at lower costs. As the technology evolves, the legal framework governing VoIP will also evolve. There will likely be a number of issues that will need to be addressed, but they are best addressed through meaningful dialogue among affected stakeholders that have a view and an incentive to create "win-win" solutions that benefit all parties. It seems that the prudent thing for Congress to do at this point is to foster that dialogue by taking a holistic approach to examining VoIP technology with an emphasis on promoting competition, preserving state authority, and protecting the public interest, rather than moving forward with a policy that preempts state taxing authority, discriminates against traditional voice communications providers, and disrupts state and local fiscal systems.

Resolution Seventeen

Preemption of State Authority to Tax

WHEREAS, the power to define the state tax system is a core element of state sovereignty, and

WHEREAS, the United States Constitution establishes appropriate bounds to the sovereignty of the states in the tax arena, and

WHEREAS, the system of federalism that is defined by the United States Constitution further cedes to state and local governments the responsibility for supplying the majority of the daily services due to its citizens and residents, and

WHEREAS, a vibrant state and local tax system is essential to meeting those needs, and

WHEREAS, the U.S. government has traditionally shown substantial deference to the tax sovereignty of the states, and

WHEREAS, there is an increasing number of groups seeking to preempt state taxation authority in particular areas, and

WHEREAS, federal preemption of state tax authority has the effect of establishing a preferred class of taxpayer and shifting the tax burden to other non-preferred taxpayers, and

WHEREAS, federal preemptions often have unintended consequences, and

WHEREAS, our system of federalism can result in substantial administrative compliance burdens for persons with tax responsibilities in multiple states, and

WHEREAS, many of the legitimate goals that might be pursued in preemptive legislation can be effectively achieved through cooperative state efforts and improved uniformity among the states, now, therefore, be it

Resolved, that the Federation of Tax Administrators respectfully urges the Congress and the U.S. federal agencies to refrain from enacting measures, taking actions or making decisions which would abrogate, disrupt or otherwise restrict states from imposing taxes that are otherwise lawful under the U.S. Constitution or from effectively administering those taxes, and be it further

[11] Walter Nagel and Ari M. Lev, "VoIP: The Second Battle of the Internet Tax Wars," State Tax Notes, June 3, 2004.

Resolved, that Congress should undertake an active program of consultation with states as it considers measures that would preempt state tax authority, and be it further

Resolved, that states should actively pursue such uniformity and simplification measures as are necessary and effective in addressing concerns of administrative burden in complying with the tax laws of multiple states.

This resolution shall automatically terminate three years after the Annual Business Meeting at which it is adopted, unless reaffirmed in the normal policy process.

Adopted at the FTA Annual Meeting, June 9, 2004

Mr. CANNON. Thank you, Mr. Cordi. We are on a remarkable roll here, where three out of three witnesses have done under 5 minutes.

Mr. KIRKLAND. I think the rule means that I get that much time to myself now. Is that——

Mr. CANNON. Well, it depends on what you say. [Laughter.]

You keep us interested, you probably have a long time. Thanks.

Mr. Kirkland, please go ahead.

TESTIMONY OF JAMES KIRKLAND, ESQ., GENERAL COUNSEL AND SENIOR VICE PRESIDENT, COVAD COMMUNICATIONS GROUP, INC.

Mr. KIRKLAND. Good morning, Chairman Cannon and Congressman Chabot. Thank you for offering me the opportunity to provide Covad's perspective on voice-over-IP and how best to ensure that this exciting new technology is rolled out as rapidly as possible.

The Judiciary Committee's oversight in this area is as important today as it has ever been, in light of recent activity in the D.C. Circuit, and frankly, companies like Covad who have invested around the 1996 act in competitive businesses, investing hundreds of millions of dollars in facilities at some point start to feel like there has been kind of a bait and switch. In January, I read the Trinko decision, where the Supreme Court discussed how the role of regulation perhaps reduced the importance of antitrust scrutiny, and 2 months later, we have a major court decision that essentially removes fundamental elements of the regulatory scheme, and we are wondering, you know, where does the buck stop?

We need antitrust enforcement. We need rigorous antitrust oversight. We also need market-opening regulation to facilitate the introduction of new competitive technologies. I think in the voice-over-IP area, this is a very exciting technology, but it is easy to get lost in terms of what it means in the marketplace and competition. And the new services that are in the marketplace, companies like Vonage and AT&T's CallVantage services, are essentially what we call applications or software. They operate on a computer, but they do not directly interrelate with the underlying broadband network.

These applications or software programs can be delivered over any kind of broadband network, and the providers who provide these services by definition do not control the underlying transmission facilities that these services ride over. They are like a Web browser or any other application that rides over the Internet. They are simply software, and the underlying transmission facilities are provided by either the phone companies like DSL, by companies like Covad over DSL, by cable companies over broadband facilities.

This is a critical point, because every time you hear about a new technology, new forms of competition, there is a big emphasis on, well, revisting the need for underlying regulation of bottleneck facilities. The local phone network remains the one ubiquitous set of loops that connects all homes and all businesses in this country. While you hear a lot about new technology, for example, the cable companies, they predominantly serve residential areas. They do not serve the small businesses of this country. All of the new technologies you hear, broadband over power line and broadband wireless, are many years away.

So for the foreseeable future, to the extent that you want innovation and competition, companies like Covad will still need to access that ubiquitous network of loops in order to provide our services. In addition, you know, the history of innovation is driven not just by the software or by the application but also by the network. The software has an area in which it can operate and function, but if you can combine innovation in software with innovation in the network, you will have a much better, more accelerated introduction of advanced features, more of a productive spiral of innovation.

And so, for example, Covad is able in the voice-over-IP arena not just to provide an excellent software package that provides all the exciting features that we have been talking about: an ability to dial phone calls off of your computer; a single inbox that has all of your voice mail, email and faxes in a single inbox; an ability to control those features, to forward calls to different numbers on the fly, so if you are going somewhere different for a weekend. And in order to do that, however, we are also able to protect the voice quality of the service that goes over that line because we control our network.

I think I would just point out, you know, voice-over-IP is here. Covad is launching the service in 100 cities. We recently had our launch party in Washington, D.C., and we expect to be nationwide by the end of this year. We raised $125 million in new capital to fund this rollout, and we are very excited about this technology, but procompetitive market regulation still has a very critical role to play.

I think one other final point is I think the history of innovation of this country shows that while large companies have been a source of innovation, small companies have been a very important source of innovation as well. So it is critical that this Committee, via its oversight as well as the legislative process preserve that competitive, those competitive alternatives, and we appreciate your attention to these issues and look forward to your questions.

[The prepared statement of Mr. Kirkland follows:]

PREPARED STATEMENT OF JAMES KIRKLAND

Good morning Chairman Cannon, Ranking Member Watt, and Members of the Subcommittee. My name is James Kirkland, and I am the General Counsel of Covad Communications. I would like to thank Chairman Cannon for convening this important hearing on VoIP services, and for allowing me the opportunity to offer Covad's perspective on ensuring the rapid rollout of VoIP. At the outset, let me also commend Chairman Sensenbrenner and Ranking Member Conyers for their foresight and leadership in promoting the rapid deployment of VoIP services through H.R. 4412.

The Judiciary Committee's oversight of the enforcement of the antitrust laws is of particular importance today in light of recent actions by a Federal court and the FCC. The D.C. Circuit's decision to vacate the primary competition-enabling rules governing access to the last mile of the telecommunications network created a vacuum which places the large monopoly phone companies in the enviable position of having a monopoly over a critical portion of the local phone network with few regulations requiring they open those lines to competitors. The FCC's efforts to fill this vacuum are critical, but unfinished. These developments, coupled with the Supreme Court's decision in *Law Offices of Curtis Trinko* v. *Verizon* limiting the applicability of the antitrust laws with regard to activities governed by the Telecommunications Act, force us to consider whether the large local phone companies now have market power to limit what would otherwise be a very vibrant VoIP marketplace. That is why it is so important for this Committee to have this hearing today, and that is why we are supportive of the Chairman and the Ranking Member's efforts to ensure that the antitrust laws and procompetitive loop access requirements continue to remain an appropriate tool to open local monopoly markets.

I would first like to discuss what Covad is doing with VoIP, then give you an overview of and some key statistics concerning the VoIP market, and finally touch on the key policy issues that are important to this Committee.

Covad and VoIP

Covad will be at the forefront of the deployment of VOIP technology. We were the first company to deploy mass market broadband DSL services in the nation, and have invested hundreds of million of dollars in building the leading nationwide facilities-based broadband network, reaching nearly 50 million homes and businesses in 35 states. Covad's broadband facilities reside in over 2000 neighborhood central offices across the nation. Today, we continue to invest in facilities-based competition. This year, Covad acquired a leading VOIP service provider, Gobeam, and in March we raised $125 million in new capital to help fund a nationwide VoIP rollout. By the end of 2004, Covad plans to roll out its business-class VoIP services nationwide to 100 major markets. In 2005, Covad will develop consumer VoIP services across its nationwide broadband facilities. As its name suggests, Voice over Internet Protocol based services bring the flexibility and capacity for rapid innovation found in other IP enabled services to public voice services. These services have traditionally relied upon the hard wired, and relatively inflexible, capabilities of the public telephone network. Covad's VOIP services illustrate the power of this combination of voice and IP. Covad's services provide businesses with all of the capabilities of expensive PBX systems, with little investment in hardware. Each user receives a unique phone number to consolidate their multiple phone numbers. Find me and follow me capabilities allow calls to find you no matter what phone you are using, and are all configurable in real time using a "Dashboard" web-interface to manage incoming and outgoing phone calls through a computer. The service includes a personal virtual fax number to handle all incoming faxes; a unified visual mailbox to manage voicemail and faxes like e-mail; and robust call logs and integration with Microsoft Outlook, allowing users to make and return calls from their PC. Covad's VoIP services also include easy to use web collaboration and voice conferencing tools. These features dramatically enhance the speed and ease with which end users can access the enhanced functionalities of VoIP telephony, combining the familiarity of a traditional telephone handset with the flexibility and power of a computer-based interface.

It is not an understatement to say that facilities-based VoIP services truly hold the potential to revolutionize the telecommunications industry, all within a few short years. Indeed, the VoIP revolution is not just around the corner—it is already underway. The U.S. VoIP market has been forecasted to grow to more than five million subscribers by 2007, a five-fold increase over 2002 levels. Furthermore, the Internet Protocol-PBX market, which has just under 100,000 lines today, is expected to grow to more than 1.7 million lines by 2007. Covad adds a unique and critical ingredient to this revolution—namely, its own nationwide, facilities-based broadband network. Covad's management of last-mile broadband transmission facilities enables it to offer VoIP services that rival the legacy public switched telephone network in their reliability, quality of service, and public safety features, such as access to 911.

The Importance of Facilities-Based VoIP Competition

Covad is able to provide innovative new services like VoIP because Congress had the vision and the foresight in 1996 to create a flexible regulatory framework to manage the transition from local telephone monopolies to robust local competition. This transition is still at a very early stage. The local telephone network remains the sole, ubiquitous public infrastructure connecting virtually every home and business in this country. By requiring that the local telephone companies allow competitors to utilize and integrate these ubiquitous loops into innovative, facilities based service platforms, competitors can develop new and innovative services like VoIP.

Vigorous innovation in the provision of telecommunications services requires that a service provider control both the "application" portion of the service it provides as well as the underlying transmission capabilities used to carry a service. By controlling its own broadband facilities, which utilize telephone company lines from a customers' premise to central offices where Covad maintains its own broadband points of presence, Covad is able to control the quality of service it provides to its customers, and introduce innovative features that are both software and network based. On the other hand, if the lines which connect homes and businesses become the exclusive province of a monopoly phone company in any area, the deployment of new technologies like VoIP will be determined by the decisions and business objectives of one, or at most two large incumbents that control facilities in any market. Covad respectfully submits that the history of innovation in this country has been driven as much, if not more, by small entrepreneurial companies as large, well funded incumbents. If VoIP is to truly flourish, there must be room for both small and large competitors. With the competitive spur of smaller, often nimbler and more focused competitors, the large incumbents are far more likely to deliver on their promises of future investment in advanced facilities.

Without robust facilities-based competition from multiple players, Covad believes that the revolutionary potential of VoIP may not be fully realized, or may be realized much more slowly. At this initial stage in the development of VoIP services, VoIP service providers that do not operate their own broadband transmission facilities have had some initial success in developing the marketplace for VoIP services. For example, in a few short years, Vonage has grown its subscriber line count to more than 100,000 consumers and small businesses across the nation.[1] AT&T recently announced its own entry into the third party VoIP marketplace, with the rollout of its CallVantage Service. AT&T plans to enter 100 major markets by year's end, and expects to sign up 1 million consumers and businesses for CallVantage services by year-end 2005.[2]

These services offer innovative features, but are limited by their providers lack of control over the facilities used to carry them. Indeed, as Banc of America Securities recently wrote,

> Because they have no legacy voice business, the virtual carriers, like Vonage, have every reason to press ahead aggressively . . . But they have significant risks long term. The current regulatory arbitrage from which they benefit (namely the ability to circumvent access charges and the USF), may go away eventually; they have little brand awareness or reputation; they can't bundle multiple services; and they are at the mercy of the infrastructure provider to maintain the plant sufficiently; and, at least today, they can't offer a quality of service (QoS) guarantee.[3]

Control over and operation of underlying broadband transmission facilities will confer significant advantages to service providers offering integrated transmission and VoIP services, such as:

> [the abilities] to control the quality of service, leverage existing customer relationships and take advantage of their on-the-ground field service networks to assist with customer installation.[4]

For example, Covad's control over its network based facilities allows it to use packet prioritization techniques to ensure that voice quality is maintained even as a user downloads large files or watches streaming media.

Competition in the underlying transmission facilities layer will become increasingly more important over time in ensuring the competitiveness of services and applications like VoIP. In other words, to preserve and extend the competition being created by third party providers of IP enabled services, it will become increasingly more important to preserve and extend competition in the underlying provision of broadband transmission services. Robust competition in the broadband transmission facilities layer for competitors like Covad who are unencumbered by legacy busi-

---

[1] *See* "Vonage Becomes First Broadband Telephony Provider To Activate 100,000 Lines," Press Release, Vonage, Feb. 2, 2004 (available at http://www.vonage.com/corporate/press—index.php?PR=2004—02—02—0).

[2] *See* "AT&T Ushers In New Era in Communication With Launch of AT&T CallVantage Service—New Jersey," Press Release, AT&T, March 29, 2004 (available at http://www.att.com/news/item/0,1847,12989,00.html).

[3] *See* "Straight Talk on VoIP," David W. Barden, et al., Banc of America Securities Equity Research, April 15, 2004, at 4.

[4] *See* "Everything Over IP," Glenn Campbell, et al., Merrill Lynch Research Report, March 12, 2004, at 19 (available at http://www.vonage.com/media/pdf/res—03—12—04.pdf).

nesses will help ensure that the exciting innovation being witnessed today in the provision of third party IP enabled services like VoIP will continue unabated.

The Market Structure

Robust facilities-based competition in the provision of the broadband services that VoIP requires does not yet exist. Amidst all the hype over the broadband future and new technologies, the underlying reality is stark. According to recent FCC data, the incumbent telephone companies and cable providers control more than 93% of the nation's broadband access lines.[5] Moreover, many end users lack a choice even amongst this limited set of two providers—for example, cable providers have historically focused their network deployment in residential areas, leaving most businesses with the incumbent telephone company as their only broadband option. In fact, recent figures show that cable penetration in the small business segment has actually dropped: "We projected cable modem would surpass DSL in this [the small business] segment by year-end 2003. However, cable modem penetration *dropped precipitously* in the small business market, or businesses with between 20 and 99 people. Cable operators also achieved limited success in the remote office market, reaching only 4.2 percent of the market in 2003."[6] As the Yankee Group now recognizes, "*DSL operators dominate* the U.S. [small business] broadband and enterprise remote-office broadband market."[7] Even more fundamentally, as both the Department of Justice and the FCC have long recognized, duopoly conditions are insufficient to produce competitive outcomes. Duopoly competition is problematic not simply because the firm with the larger market share may exercise market power, but also because *both* participants are likely to have the incentive and ability to maintain prices above competitive levels rather than attempting to ruthlessly compete with each other, as they would need to do in a market with multiple firms.[8] Accordingly, as the FCC has concluded, "both economic theory and empirical studies" indicate that "five or more relatively equally sized firms" are necessary to achieve a "level of market performance comparable to a fragmented, structurally competitive market."[9] Most importantly, large incumbents with substantial investments in existing facilities are less likely, left to their own devices, to be aggressive innovators in disruptive technologies like VOIP.

The incumbent telephone companies, with substantial legacy businesses, face conflicting incentives in deploying VoIP, which threatens their core circuit-switched voice businesses with VoIP services:

> SIP threatens to strand the Bells' core network . . . VoIP customers bypass, obsolete and strand the Public Switched Telecom Network (PSTN).[10]

Given nearly $150 billion invested in circuit-switched telephone plant,[11] it is easy to see why incumbent telephone companies have severely conflicting incentives in rolling out VoIP: "the Bells will be reluctant to cannibalize themselves . . ."[12] The Bells' history in deploying DSL technology is instructive. As is now widely acknowledged, the incumbent phone monopolies were slow to deploy ADSL precisely because it threatened to cannibalize lucrative, legacy monopoly services such as ISDN, T1, and second line telephone service.

The cable industry also has conflicting incentives. Cable providers have much stronger incentives to aggressively roll-out bundles of VoIP and broadband transmission. After all, "[r]elative to the Bells, [cable's] major advantage is obviously that it doesn't have a legacy voice business it needs to protect."[13] Viewed in the broader context of their own legacy monopoly, however, the picture gets murkier. Under du-

---

[5] *See High-Speed Services for Internet Access: Status as of June 30, 2003,* Industry Analysis and Technology Division of the Wireline Competition Bureau, Federal Communications Commission, at Table 5 (December 2003). Specifically, out of a total of 23,459,671 high-speed lines (over 200kbps in at least one direction), RBOCs served 7,266,765 lines, other ILECs served 948,828 lines, and cable providers served 13,684,225 lines.

[6] Yankee Group, *Cable and DSL Battle for Broadband Dominance* (February 2004), at 4–5 (emphasis added).

[7] *Id.* at 4 (emphasis added).

[8] *See* United States Department of Justice/Federal Trade Commission, *Horizontal Merger Guidelines,* Section 2 (rev. Apr. 8, 1997).

[9] Report and Order, *2002 Biennial Regulatory Review—Review of the Commission's Broadcast Ownership Rules and Other Rules Adopted Pursuant to Section 202 of the Telecommunications Act of 1996,* 18 FCC Rcd. 13620, ¶289 (2003).

[10] *See* "SIP Happens: How VoIP Technology 'Re-unbundles' Telecom," Scott Cleland, et al., Precursor Telecom and Media Research, Apr. 12, 2004.

[11] *See id.*

[12] *See* "Straight Talk on VoIP," *supra* n. 3, at 4.

[13] *See* "Straight Talk on VoIP," *supra n. 3, at 5.*

opoly conditions, the ILECs and cable providers have every incentive not to aggressively compete in each others' core businesses:

> [W]e think cable operators are wary of being too successful . . . the chief risk is that being too successful in VoIP could induce the Bells to be more aggressive in the data and video businesses (such as ratcheting up marketing activity and price pressure). To put it another way, we think cable operators want to be successful with VoIP only up to the Bells' threshold of pain; maximizing the value of VoIP may not maximize the value of the cable business if it invokes a predatory response . . .[14]

> [W]e think cable regards the potential Bell threat as much larger [than virtual carriers like Vonage] and we think it is highly unlikely to risk baiting the Bells with an aggressive push into VoIP just to preempt what it regards as a smaller threat.[15]

Indeed, alongside the flurry of press announcements announcing cable operators' ambitious future VoIP rollout plans is a note of caution:

> Most are wary of using big, new capital expenditures to take on entrenched local phone giants, such as Verizon, while they are also spending heavily on fancy, new set-top boxes and cable modems. "To dislodge a competitor that large takes a lot of money, and cable operators are still loaded with debt," says Richard Nespola, CEO of telecom consultant TMNG. "Investors would not jump for joy."[16]

This economic reality highlights another limitation of duopoly competition in the IP transmission layer. To the extent that the cable industry does pursue VOIP services, this is no guarantee that the industry will make further investments to optimize their transmission networks for VOIP. They may merely elect to provide VOIP services on a "best efforts" basis utilizing their existing internet access capabilities. In this scenario, cable companies would not drive any significant transmission layer innovation, but would simply be "virtual" voice carriers, like Vonage, over their own networks.

Unlike the established telephone and cable companies, Covad and other competitors have no legacy business to protect. Thus, we believe that including Covad's facilities-based VoIP offerings in the overall marketplace will significantly speed the rate at which broadband services like VoIP are adopted, and the development of innovations in these services.

Lessons from Abroad

The experiences of countries like South Korea and Japan are instructive. Both nations enjoy significant leads over the U.S. in broadband penetration, and both nations have experienced explosive growth in broadband deployment after adopting and enforcing unbundling regimes. South Korea's market-opening measures included the formation of a new company (Hanero) to compete with incumbent Korea Telecom,[17] and opening Korea Telecom's network with requirements for local loop unbundling, including sharing of the local loop.[18] The result has been thriving competition in the broadband market, with three main suppliers,[19] and rock-bottom prices (as low as $25 a month[20]) for consumers. As a result, "[a]t the end of June 2003, South Korea ranked third in the world by the total number of DSL lines and first in the world in terms of DSL penetration, with 14.27 DSL lines per 100 population."[21]

Japan's market-opening measure included being one of the first countries to introduce line sharing, reducing line sharing charges to the lowest rates in the world, reducing collocation costs, shortening provisioning intervals, and unbundling backhaul facilities.[22] As a result of such actions, at the end of 2003, Japan led the

---

14 *See id.*
15 *See id.* at 6.
16 *See* "Cable Poised to Offer Phone Service—Just Not So Fast," USA Today, May 27, 2004.
17 *Id.*
18 *See* "Developments in Local Loop Unbundling," Organisation for Economic Cooperation and Development, Working Party on Telecommunications and Information Services Policies, Sept. 10, 2003, at 49 (available at http://www.oecd.org/dataoecd/25/24/6869228.pdf).
19 *Korea Broadband,* PDS Consulting Short Paper, Version 12 June 2003.
20 *Seoul's Strong Hand Sets Pace on Web,* International Herald Tribune Online, November 26, 2001.
21 *South Korea,* Korea Broadband Overview, Point Topic, October 20, 2003.
22 *On a roll: Japan's success with DSL,* Ovum Research, DSL: Business Models for Exploiting the Local Loop, July 2002.

U.S. in broadband penetration, and a competitor named Softbank—not the incumbent—was the top DSL carrier in Japan.[23] The experiences of South Korea and Japan show that maintaining competitive access to local loop and transport facilities spurs the deployment and adoption of innovative new services like broadband. Similarly, preserving competition among multiple facilities-based providers of VoIP will dramatically speed the pace at which VoIP services are developed, deployed and adopted here in the U.S.

VoIP Policy Issues

Aside from minimal regulation ensuring access to the last mile of the phone network, we believe that policy makers should adopt a generally deregulatory stance towards VoIP. We believe there is promising evidence that traditional social policy objectives can be met without enacting new regulatory requirements for VoIP services. Of particular importance to this Committee is law enforcement access to communications conducted over IP enabled services. First and foremost, I can tell you that Covad is committed to working with all law enforcement agencies to ensure that those officials have access to all the information from a VoIP call that they currently have access to for a regular phone call. In fact, we have complied with such requests in the past. In addition, last December, the National Emergency Number Association (NENA) and the Voice on the NET (VON) Coalition, of which Covad is a member, announced a voluntary agreement on approaches to provide VoIP subscribers with basic 911 service, and to work together to develop solutions for enhanced 911 functionality.

Furthermore, we believe that many critical social policy objectives can be met by focusing on enforcing and rationalizing existing telecommunications service regulations, rather than by extending them to information services like VoIP. For example, we generally believe that regulators should refrain from imposing legacy access charge regulations on VoIP services, and instead should focus their efforts on reforming existing regulations to develop a comprehensive intercarrier compensation mechanism. Similarly, rather than imposing new universal service obligations on information services like VoIP, we believe that regulators can help safeguard universal service by rationalizing the existing contribution mechanism, so that all providers of broadband transmission services contribute equitably. In sum, we believe that the enforcement of existing regulations on broadband telecommunications service providers like Covad, combined with voluntary industry collaborative efforts and standards setting, can meet critical social policy objectives like public safety and universal service—without imposing intrusive new forms of regulation on information services like VoIP.

Conclusion

Mr. Chairman, Members of the Subcommittee, we are in the midst of a revolution in the telecommunications industry. We are moving away from the limitations of traditional phone service towards all of the enhancements, efficiency gains and innovation that VoIP makes possible. We are moving away from competition through legacy circuit switches to facilities-based competition over packet-switched broadband networks. Because of all that, now more than ever this Committee's oversight and stewardship of the antitrust laws is crucial. I hope that we can work with you in the future on these very important issues.

Thank you again for this opportunity and I welcome questions from the panel.

Mr. CANNON. Thank you, Mr. Kirkland.

The Chair recognizes the gentleman from Ohio, Mr. Chabot, for 5 minutes.

Mr. CHABOT. Thank you, Mr. Chairman.

Unfortunately, I have a flight in less than an hour, and security being what it is these days, one never knows how long it is going to take to get through security. So I will yield my time to the Chairman to grill the witnesses here this morning, and I want to thank them for their very interesting and informative testimony, and my staffer is here as well, so we will be following very closely and look forward to working with all of you in the future on this important technology.

---

[23] *How the "Japanese Miracle" of Broadband Came About,* Glocom Platform, Japanese Institute of Global Communication, Colloquim #43, December 24, 2003.

Thank you. I yield to the Chair.

Mr. CANNON. Thank you. I thank the gentleman and appreciate your being here today to help us getting started.

I want to apologize to the minority, which is not here. Both Mr. Watt and Mr. Delahunt asked that we defer the hearing. Both recognized the commitments by the members of the panel, and since Mr. Kirkland had already embarked from California to arrive, we suggested that we would go forward with the hearing. And we will try and keep the interests of all parties in mind as we ask some questions. Actually, the "we" is not royal. The "we" is actually me, I think, here today. So I appreciate your attendance, and I know that is at some sacrifice coming from across the country. I appreciate that, Mr. Kirkland, and thank you for your testimony.

You know, Mr. Kirkland, you just mentioned the issue of small companies and how they relate here, and I think that is actually one of the most interesting issues before us. My district has a huge amount of information technology, and having the rules clear on VoIP is important. So you have a few genius type guys who with some few thousand lines of code can come up with an entirely new product or concept that transforms the world.

If you have VoIP available, it seems to me that is important. I would actually like your thoughts on that Mr. Kirkland and also Mr. Cordi, but in addition, if I could just point out that we had a company—I think it was in Washington, yes, the Washington State regulators found that VoIP provided by a local company called Local Dial was a telecommunications service. This was a very tiny company and then ordered Local Dial to register and comply as such, which included the remission of access charges.

About a week later, Local Dial shut down, because it concluded it could not comply with the order and stay in business. Is this not a clear demonstration of the destructive power of taxation and that in an environment where we want to create a fertile field for innovation, taxation in this new area may actually be deathly?

And Mr. Kirkland, do you want to comment and then Mr. Cordi?

Mr. KIRKLAND. Yes, this is why we support the approach taken in the Pickering legislation. We do believe that a very light regulatory touch on voice-over-IP is critical. This technology is very exciting, but it is very much in a nascent stage, and the kinds of entrepreneurs that you described as well as even larger companies would struggle with a 50-State regulatory regime over voice-over-IP. So we are very supportive of that approach. We are also supportive of a very light touch with the caveat that I discussed in my testimony, and that is structural regulation of the telecommunications market remains critical, and in fact, it will enhance the rollout of these technologies.

So we are very sympathetic, and even with larger companies, there are substantial costs involved in complying with the whole myriad of State regimes, and so we do think the Federal level is the appropriate level for policy here.

Mr. CORDI. Mr. Chairman, on the tax question, we certainly do not deny that compliance with State tax requirements and local tax requirements presents some burden. There is no question about that. But we do not think that the first reaction of the Congress should be because of that burden to preempt State taxation in its entirety. It is clear to us that to create this preemption creates an unlevel playing field which threatens the remainder of a very large source of State and local revenue.

The States and taxpayers have proven that they can work together to address burden problems, and I think we feel strongly that we should be given a chance to do so.

Mr. CANNON. Can I just follow up with that, Mr. Cordi, for a moment, and ask you to help balance for me the burdens. Let me say it this way: you have this new technology. Mr. Pepper referred to it as comparing VoIP and traditional telephony with a garage sale and an Ebay sale, and I think that, literally, the magnitude of difference is that much. Maybe the same thing is that you have stuff in your garage you want to sell; one is more efficient.

There is another element here of differentiation, which is that we cannot even imagine the kind of tools, the kinds of products that may become available as people look at this. So those products, in my experience, and I have—we have had some large IT companies, and one of the funniest things I have ever watched in my lifetime is we had a fellow named Ray Norda who ran one of our great companies, and he had the view that he should fire 10 percent of his people every year. And so, every year, he would have a 10 percent layoff, and these guys would all go out, and they would say, well, I have got three offers from big IT companies, and I have five buddies who each have a new IT idea that they are working on, and before the bubble, of course, that was a lot more attractive.

I will say that most of those guys are back to work now, which is very nice. But I have lived with, and I actually did venture capital with some of these companies. So the amazingness of some of the ideas is what I think we ought to be aware of in the future. But, you know, there are all kinds of problems with a start-up. In the first place, if it is really a good idea, and it is really going to threaten the establishment, it gets absorbed pretty quickly, and the world changes. And, of course, the major companies in America have proved that they can be adaptive, led, by the way, by AT&T, which did an audacious thing 5 years ago or 6 years ago to enter this space.

So what I would like you to do, Mr. Cordi, is to sort of respond. I understand that this is a source of cash, and in fact, if I might just go a little bit further, I heard an estimate the other day that the cost of switching, the cost of providing a phone call over the Internet, a VoIP phone call, is less than one-fifteenth of a switched call, and I think it is probably significantly less than that, and the scaling makes it even less.

But in a context where you have a shrinking cost base, if your taxation stays at a relatively constant percentage, your revenues are going to shrink anyway. As those revenues shrink, as we are in a market where new ideas emerge that make the world a better place and which drive the whole economy, because but for the last couple of years when we have had a little bit of a slowdown, the information technologies have driven State revenues at a remarkable pace.

Is there not a reason for the States to back off and say we probably ought to let this grow?

Mr. CORDI. Well, I guess first of all, the number we are talking about here is not a small number, because we perceive this as threatening the whole source of revenue.

Mr. CANNON. Now, when you say the whole source of revenue, you mean the telecommunications taxation.

Mr. CORDI. In general; certainly, the money we take in from landline phone services and from this. And it gets worse to the extent that this preemption would facilitate——

Mr. CANNON. Right.

Mr. CORDI.—the move of business in that direction.

Mr. CANNON. That was a long question, so let me break it down in pieces. As you look at a reduced cost of services, you either have to expand the rate of taxation, or your revenues are going to fall. Is that not a concern?

Mr. CORDI. Well, our taxes are based on the charges in general, not on the costs. And so——

Mr. CANNON. Well, if it is a percentage of the charges by the phone company, except for some of the fixed costs; there are some fixed taxes, and there are some percentage taxes. If the cost declines, that is, if the cost of providing the service declines, and you are in a highly competitive environment, which we are, you are going to see a decline in the cost or in the charges that the phone companies make and therefore a decline in revenues.

Mr. CORDI. Revenues will go down, Congressman. You are quite right. And that will present a problem for local policy makers. And to the extent that they need that revenue, they are either going to have to increase rates for their services or find other sources of revenue or cut expenditures. I do not think there is any other alternative. I am sorry; you were about to say something.

Mr. CANNON. Let me just go a little bit, a step further. I met with AT&T recently to see their VoIP product in anticipation of this hearing, and the woman who made the presentation said this is a $34.95 price, but for the first 6 months, it is $19. So, I said, well, does the $19 fee require some long-term contract, or if prices decline in the future, you know, and I sign up, am I going to have a reduced price?

And there was some confusion, and finally, one of the guys said look: we are in a market where prices are declining. You will be lucky to maintain that $19 price. So the introductory hook price is likely to be the high end of the long-term price, and you are talking about a service that retails for $34 but is selling for $19 and is going to fall to $8 or whatever I would like. You set your prices, not me.

So in that environment, you are looking, and now, of course, the QWest has a naked DSL, meaning you can do just DSL without a line. That means—I use QWest at home. My bill recently went from $150 to $75 with virtually all of the same services. I cannot get DSL where I am right now, although I think that is coming in the near future. When I get DSL, I will be able to have a line charge—they have got two, now, standards. One is 512K, I think the other is 1.7 meg.

So for the same price as one line today, which is not taxed, by the way, because currently, at least, Utah is not taxing, I will get DSL service, and then, for $19 or some other amount of dollars. So for less than two-thirds of what I am paying right now, I am going to have all of the telephony that I can use, because I think that includes long distance if I use AT&T's product; I am not sure what the QWest product is. I know that my bill is going to go down in the future, and even my underlying DSL service is going to go down as other competitive services come on board.

So I am looking at a reduction in my phone bill today of a third and probably a reduction to about half or less over the next year or two or three. That means your revenue base is declining like crazy, and your constituents are not going to let you keep that cost up, do you think?

Mr. CORDI. Well, I think you are completely correct. The likelihood is that revenues may come down here, and that will present a revenue problem for State governments. But I would argue we do not want to aggravate that. That is going to happen regardless of what you do here. We do not want to aggravate that by creating a preemption that sort of takes all of it out of the picture and over a short period of time.

Getting back to the underlying problem, which is dealing with the burden question, I mean, what we see here is the big players, AT&T already, you know, pay taxes around the country. They have existing systems to do that. The burden here will be incremental. The new players, for the most part, will not be subject to our reach because of nexus questions, that we will not have the authority to reach a player that is out in California in Maryland unless he has got some presence there.

And so, I see for the startup people a period of time during which they are not going to have this tax obligation until they become more present or unless Congress passes something like the streamlined sales tax which would provide a tax payment requirement without regard to nexus.

Mr. CANNON. This is a real complicated issue, and I really actually want to hear from our other panel members. But I am not letting you off the hot seat, because this is the dialogue that we need, that is really important. And I apologize for giving such long questions, but the context, I think, is important.

And now, you have touched on several different things. If I might just deal with SSTP for a moment, the streamlined sales tax program, it seems to me that we just got a letter from, which we will make part of the record without objection——

[The information referred to follows:]

National Governors Association
Council of State Governments
National League of Cities
The U.S. Conference of Mayors
National Association of Counties

July 19, 2004

The Honorable F. James Sensenbrenner, Jr.
Chairman
Committee on Judiciary
U.S. House of Representatives
Washington, D.C. 20515

The Honorable John Conyers Jr.
Ranking Member
Committee on Judiciary
U.S. House of Representatives
Washington, D.C. 20515

Dear Chairman Sensenbrenner and Representative Conyers:

We urge you not to take action on H.R. 4129, the "VOIP Regulatory Freedom Act of 2004" this year. While the objectives of H.R. 4129 may be well-intended, the bill is a premature response to an emerging technology that fails to adequately address the wide variety of complex issues that face the communications and Internet industries as a whole.

Voice over the Internet technologies (VOIP) and other advanced communications have the potential to provide consumers with more choices among phone services. However, determining the appropriate regulatory treatment for these new technologies will require policymakers to look at these issues holistically, and ideally, develop policy that does not create different rules for different technologies. For example, given predictions that VOIP may replace conventional telephone service in the future, it is important that the state and local government roles promoting the public interest in traditional telecommunications apply equally to VOIP. Unfortunately, H.R. 4129 would preempt virtually all state authority over IP-enabled technology, undermining the roles of state and local government. Specifically, state and local government officials oppose H.R. 4129 because:

- Public safety concerns are a priority for state and local governments, yet H.R. 4129's preemption of state and local authority over E911 services and wiretapping laws would undermine state and local public safety programs.
- The elimination of state and local government from the regulation of VOIP also eliminates the ability of states to provide consumer protections, including access requirements for individuals with disabilities, or promote long-term universal service goals.
- By focusing on one particular technology, H.R. 4129 would artificially create winners and losers within the communications industry, undermining the objective of fair and open competition.
- The preemption of state taxing authority over VOIP services unnecessarily interferes with state sovereignty over state and local revenue issues.

- Finally, unless changes to communications policy are aimed at balancing the playing field for all participants—both new entrants and current players—and take the public interest fully into account, the fast-paced nature of technological developments will ensure that any new technology-specific regulatory structure will become obsolete in a relatively brief period of time.

We believe the Committee needs to take a comprehensive look at all issues facing the communications industry, a look that H.R. 4129 does not afford. History has shown states and local governments to be good stewards in the effort to promote competition and protect the public interest as it relates to the communications industry. While communications platforms may continue to change, the public interest will remain, and state and local governments should continue to be involved in protecting that interest. Thus, we urge the Committee to avoid a rush to judgment this year by not acting on H.R. 4129.

Thank you for your consideration.

Sincerely,

Raymond C. Scheppach
Executive Director
National Governors Association

Donald J. Borut
Executive Director
National League of Cities

Daniel M. Sprague
Executive Director
Council of State Governments

J. Thomas Cochran
Executive Director
The U.S. Conference of Mayors

Larry E. Naake
Executive Director
National Association of Counties

Mr. CANNON.—the National Governors Association and the National Council of State Governments, the National League of Cities, the U.S. Conference of Mayors and the National Association of Counties which pretty much lays out in brief your main points.

What we have, I believe here, if I can just make a statement, is an attempt by people who live in the status quo to strangle the emerging future, which is better for us all. I liken it to the golden goose. It is laying eggs of great value to society and to the States and to State revenues in particular. And I might say that it has implications for the rest of the world.

To the degree that we scale up, it makes it easier for people in other parts of the world to get these services. We are now talking, a group of us are trying to work with Haiti to get a WiFi system on the cheap there that would allow people there to change their lives dramatically by having medical resources they do not currently have by having agronomists help them with their crops by having a market like Ebay's market to sell their products.

You know, the biggest employer in Afghanistan today is a Utah company called Overstock.com. They employ the largest number of people and the largest number of women. So we have a bunch of women who have made carpets for their whole lives now make their carpets and sell them directly on Ebay. Overstock.com creates a context where they assure quality and delivery, and you have made the world a dramatically better place in Afghanistan.

So the issue here is not what happens in Maryland or in Utah so much as it is what happens throughout the whole world. And I cannot overemphasize the fact that the tools that we can make available very cheaply like Overstock.com has done are much more important in the long run than the soldiers who risk their lives day-to-day there in the parts of the world that are unstable.

So the transformation that we are dealing with here, the discussion that we are having about VoIP is not a discussion about the tax health of any particular State or one industry over another or one technology over another but the health of the world in a very real sense. That said, by way of admonition and maybe by counsel for you, it seems to me that if I were in the State's position, I would be saying we have the telephone revenues and the Internet tax moratorium. We have got the SSTP and what that provides for us, and then, we have got the business activity tax. And those three things combined represent the future of taxation by the States.

And to resist mightily on the Internet tax moratorium seems to me to be counterproductive for the other two. And I think that Mr. Delahunt, who serves on this panel as well and who is the leading minority pusher of the SSTP agrees entirely with me on the subject.

First of all, am I right about the relationship between those three different taxes and the future, and secondly, is there a way that we can get the various groups together so that we can come up with a rational decision instead of strangling the baby as it is born?

Mr. CORDI. We need to all be talking, you know. The State governments disagree vociferously that the business activity tax should be related to anything else, and we do not see that as a reasonable price for either the streamlined sales tax legislation or, for that matter, an acceptable Internet tax freedom act. We see that as simply unrelated. You know, for the most—for many States, the cost of that exceeds any conceivable benefit from the streamlined sales tax. So State governments, I believe, resist the linking that you have suggested.

Mr. CANNON. Perhaps we can come back to that, but stepping back, am I correct about the importance of these new technologies? And should there be a relationship between the Internet tax moratorium and the SSTP? In other words, could the States give up the potential revenues that are going to decline anyway in the context of improved revenues through the SSTP?

Mr. CORDI. Well, we are talking on these telecommunications taxes upwards of $10 billion. I do not have off the top of my head what the conceivable numbers are on the streamlined sales tax, but I am not sure they are in that range. Forgive me, Congressman, for not knowing that off the top of my head.

Mr. CANNON. You know, I have seen lots of different numbers. Business Week had a number about a month ago of $35 billion lost to the States through sales over the Internet. That seemed a little high to me, but that is one of the numbers that is out there.

Mr. CORDI. It seems very high to me, and as you know, the streamlined sales tax legislation has thresholds in it that really take out a lot of the potential revenues. And so, I do not think the number is anywhere like that, Congressman, but I do not have the numbers in front of me.

Mr. CANNON. That is right, but, you know, the MTC number which we are talking about, the $10 billion, I think, came mostly from the MTC, the Multistate Tax Commission, represents a number that we have already agreed here, I think, is going to decline significantly just because the charges that are made to the customer are going to decline. So it is not $10 billion versus some portion of $35 billion; it is a shrinking $10 billion against a growing other source. And so what I am asking is, is the question I am asking relevant to the States?

Mr. CORDI. I think the question deals with the Internet tax freedom act and the streamlined sales tax. The answer is—yes, although there is not a whole lot of overlap between the two proposals.

Mr. CANNON. That is right.

Mr. CORDI. There is some, but, you know, you can discuss the two of those separately.

Mr. CANNON. That is true, but, of course, the States have held up our version of the Internet tax in the Senate, and I think they did that without a lot of thought. What I am wondering is is there a possibility of getting the folks together that are actually thinking about this and changing the paradigm among the States?

Mr. CORDI. I think that probably, the Federation of Tax Administrators is not the key player here.

Mr. CANNON. Right.

Mr. CORDI. I think you need to be dealing with the National Governors Association and the NCSL and the other senior——

Mr. CANNON. There are other players. You guys are sort of the——

Mr. CORDI. Humble tax collectors.

Mr. CANNON.—smart guys, though, with all due respect, and I am hoping that you will take up the burden.

Mr. CORDI. Yes, sir.

Mr. CANNON. Let me just point out: there is a difference between the BAT. The reason I raise them in the same context is because to the degree that the States need revenues, they need clarity of rules, so that, in other words, I am not using the BAT to bat the States over the head but rather to say we need to have clarity about how revenues are generated so that business can operate in an environment that is predictable, and that seems to me to be the major connection there.

Let me shift here. Thanks, Mr. Cordi. I appreciate this. You know, this is a real difficult topic, and it is difficult in large part because of the fundamental transformative nature of what we are dealing with here.

And so, Mr. Kirkland, if I could just ask a couple of questions of you, how many companies do you know of that are doing VoIP, and can you give me a sense of the size? You have the monsters, but you also have the small companies and the real startups.

Mr. KIRKLAND. A lot of companies have talked about doing voice-over-IP. AT&T, obviously, showed some leadership in the space. Vonage is another company that has a lot of voice-over-IP customers. We acquired a company called Go Beam that focuses on the small and medium-sized business, and they were venture-backed and running, you know, basically trying to raise their next round, and we are now taking their product and launching it nationwide.

When we bought it, they had about 13,000 line equivalents. There are—it runs the gamut. I do not know if you read—there was an article, I think, in the Wall Street Journal this morning about a company called Skype that basically just allows downloadable software so you can make free calls over the Internet so long as the person on the other side has the same software on their computer.

So there is a whole range of companies providing these services. I think in the aggregate, it is still probably less than 0.3, 0.1 percent of the total number of communications lines out there, so it really is a nascent technology. But that is what is great about it. There are probably companies that none of us have heard of here that are providing the service and a lot of diversity out there.

Mr. CANNON. Let me ask a question, Mr. Kirkland, of you, and Mr. Langhauser and others may want to comment on this as well. I think, Mr. Langhauser, that you announced yesterday that you are leaving the local market. So you are facing some pretty significant transition in your business. You mentioned, probably when we were talking beforehand that probably about only 20 percent of the homes in America have broadband. But you pass more than 80 percent of the houses in America, as I understand it; is that not correct, with your broadband services?

Mr. LANGHAUSER. Actually, we offer broadband connectivity only in partnership with other companies, including Covad.

Mr. CANNON. Yes; thank you.

But Covad, Mr. Kirkland, Covad passes, with your partnerships, with QWest, with AT&T, how many homes do you pass in America?

Mr. KIRKLAND. Our network, as we said, we are a facilities-based company. We actually have our own facilities in 2,000 central offices throughout the country. All we use are those local loops to connect to our own DSL equipment. We pass about 50 million homes and businesses in the United States, so that is approximately half the country; generally the top 100 markets.

Mr. CANNON. My sense is that about between cable and DSL, 85 or 90 percent of the homes have access to if they do not use broadband; is that right, Mr. Pepper?

Mr. PEPPER. That is right. We estimate—it is hard to know precisely, but we estimate between 80 and 90 percent, 85 and 90 percent of households have broadband available to them through either their cable company or through DSL, trough the incumbent carriers and providers or competitors like Covad.

Based upon the latest numbers that we have seen in terms of industry reports, about 25 percent of American households now subscribe to some form of always-on, high-speed Internet service. And we also believe that some of the more exciting new technologies to provide broadband, especially in rural areas, are with wireless networks. We estimate that there is between 1,500 and 2,000 small wireless Internet service providers, many of them using unlicensed bands and unlicensed devices to provide broadband in rural communities that do not have DSL or cable modem service available.

Mr. CANNON. And those wireless services are broad enough bandwidth to support VoIP?

Mr. PEPPER. Yes.

Mr. CANNON. Mr. Langhauser, you talked about VoIP being the killer app. What does that mean in the market? I mean, if you have all of these people who have access who have chosen not to take broadband because of the cost, because they do not get the benefit, what does it mean? And may I ask also, we have had a lot of confusion in pricing. QWest's price for very narrow broadband was up to $70, $69.95 for a significant period of time. That was not the kind of thing that anybody except the real geeks wanted.

As the uncertainty settles out, as prices fall, will prices fall, and will the cost of VoIP services fall, and what will that do to the market, in your estimation?

Mr. LANGHAUSER. What I mean by VoIP possibly becoming the killer app is, as you point out, houses are passed by broadband, but for a number of reasons, consumers have not subscribed in overwhelming numbers. It is about 25 percent. And they need a reason to pay the $30, $40 a month for broadband. Some people are reluctant to use it for narrow band email, and it may not be useful for narrow band email.

This may be the application, especially as we add to it and enhance it that gives consumers a reason to have that broadband connection into their house. What is exciting about this service are some of the applications that you can put on top of the voice traffic. Mr. Kirkland mentioned some of the features. There are going to be more. And these are going to provide a real opportunity for entrepreneurial companies to help us develop features that we could put on our service.

The VoIP pricing so far has been extraordinarily competitive, almost frighteningly competitive for a service that is just being rolled out, and I expect it will continue that way, and competition tends to lower prices.

Mr. CANNON. I think of thrilling as opposed to frightening, but I am on the other side of the equation.

You spoke earlier, Mr. Langhauser, about taxing by phone number. Now, at this point, I am pretty anxious not to see any taxes go any way, and so, you can you a little bit of opposition there, but does that not have some inherent problems? For instance, my understanding is that most VoIP services, I can get an area code where I want it. You know, if my mother lives in Utah, and I am out here, I can use a Utah area code so she can call me directly, or if I want the status of a Manhattan area code, I can do that as well.

And by the way, I live in two places, and many people have different places that they locate. Does that create a problem in your mind?

Mr. LANGHAUSER. I think the fact that voice over the Internet does not comply with any of the traditional jurisdictional notions certainly causes a problem on State taxation. And you are absolutely right. You could take your Washington VoIP number to Utah, and you would have a real issue of which jurisdiction taxes that.

What I was referring to, though, was our proposal to reform the Federal universal service fund. Right now, that fund is funded only through interstate telecommunications revenues. It is a very narrow base. It is a shrinking base. It is a fund that is headed for severe problems. And what we are suggesting rather than tack on additional services like VoIP to this broken fund that the FCC should fundamentally reform it.

They should probably base the charge on telephone numbers or other connections to the Internet; subject all telephone numbers to a monthly charge. It would include VoIP; it would include wireless; it would be nondiscriminatory, and it would also sustain the life of the fund.

Mr. CANNON. So you are only thinking about the universal service fund when you think about that.

Mr. LANGHAUSER. That is right.

Mr. CANNON. But I think the States are going to have something to say about that.

Mr. Cordi, do you want to respond to the difficulty that represents or the opportunity?

Mr. CORDI. Well, I am not certain that I have anything to add to that.

Mr. CANNON. I am just concerned here about if people, if we tax, if we create or if we use the phone number as the basis of taxation, how do States participate in that process? How do they get a revenue stream?

Mr. CORDI. Well, phone bills generally are controlled by the billing address of the customer and not by the area code he happens to be in. You know, there is good precedent for collecting tax on telephone services, not only the Mobile Telecommunications Sourcing Act but elsewhere for taxing phone services at the principal place of use. And typically, if you cannot identify it to any other place, it is the billing address.

Now, that is something that even an Internet provider, anyone who takes a credit card over the Internet asks for an address. And so, even if you are billing to a credit card and not, maybe, like AT&T, sending a bill to someone's home, I think this is a manageable problem.

Mr. CANNON. I worry at some point that if we have a regime that is based on addresses or billing addresses that people will be driven to the jurisdiction with the lowest taxes, which is part of the reason that I think we need a rational solution for all States as opposed to competing interests.

Let me come back to USF, and this is a question, Mr. Pepper; you may want to talk about this; Mr. Langhauser, you may want to as well or Mr. Cordi, Mr. Kirkland. The fact is the costs of using the VoIP are much lower than the costs of switched telephony. Does that lower cost not have significant implications for the need for the USF fund? You said you have serious problems coming, but if you can use a lower-cost system, is that not actually helpful for the USF?

Mr. PEPPER. Well, I think that this is why we have some optimism. Number one, affordable phone service is a goal shared by, you know, the FCC, State commissions, Congress, State legislatures, everybody. So the goal of affordable phone service does not change. What will need to change, as you have been pointing out, is the way in which we achieve it in this new world.

In a world in which the costs are actually lower, right, it makes it easier to achieve the affordability goal. So if the costs are lower, prices can be lower, and it will be easier to provide affordable phone service to everybody. So I think you are absolutely correct that there are some significant advantages using not just voice-over-IP but other new technologies to provide the physical transmission connection as well as the applications like voice-over-IP. And again, I think back and look at some of the wireless broadband providers that are providing services to, broadband services to communities that do not have any other broadband choice. And 2 years ago, we were told those communities would never have broadband. Today, they have broadband service, and it is being provided by people with no subsidies.

Mr. CANNON. Exactly; thank you very much for that comment.

Let me go back to just one point you made and flesh that out a little bit. You talked about affordability, and this is for the whole panel, not just for you, Mr. Pepper, but affordability. Is not a tax on a fundamental service the most regressive tax you can have? In other words, as you think about that for just a moment, John Conyers and I, the Ranking Member of the full Committee, have had a long association in this particular battle, because the digital divide leaves people that he believes he is representing on the wrong side.

And so, we have worked strongly together to try to help bridge that digital divide. To the degree that we are taxing these kinds of services, is that not extraordinarily regressive, and Mr. Pepper, I would like you to respond first. You seem to be interested, Mr. Langhauser, as well, but we will let you do the cleanup, Mr. Cordi, and give the other argument.

Mr. PEPPER. I am not a tax expert, but you are absolutely right that, you know, people at the bottom end of the economic scale can least afford to pay more for services, and one of the ironies that we have seen is that many of the universal service or other fee or tax obligations tend to fall very heavily on low-income people, especially low-income people who make a lot of long-distance calls. And that really also applies to immigrants.

So we have a lot of, you know, people who come to the United States. You know, it is still the country of everybody's dreams because of economic opportunity. They do not make a lot of money. They call family members back home. They are paying very high fees that actually increase their costs, and so, in some ways, it does not really help close that gap.

Mr. LANGHAUSER. I think one thing I would add to that is for some reason, and it is probably history, telecommunications in general has been singled out for a myriad of different State and local taxes. In some respects, it is treated as if it were tobacco or alcohol, with almost a punitive tax burden. I think this is something that is very important to address, and we are not arguing that we should not be taxed at all. We are arguing that we should be taxed like regular businesses and not singled out for excessive tax burdens.

Mr. KIRKLAND. To build on what John said, I think you also see in the various taxes and fees, as you often see in communications issues, real inequities in what kinds of services, even services that appear like like services, some contribute; some do not. You know, USF is a good example, where cable modem service does not contribute into USF; other forms, like DSL, do in certain circumstances.

And so, there is a whole legacy set of different fees, taxes, other sorts of things that the current system needs some rationalization. And before you then take some exciting new technology which certainly has great potential but extend, you know, systems that are in need of sort of a fundamental re-look or fundamental reform, another example being access charges, you know, we would suggest that you do not just take the old legacy system and try to figure out where to pigeonhole voice-over-IP but really look at the fundamental premises of these.

And that is not to say that voice-over-IP should not bear its fair share, but there is some fundamental restructuring and equity that needs to be brought to the process. And I would argue that while this technology is nascent, while it is evolving, while it is still developing, and we will see where it ultimately ends up and what it really looks like on the ground, because really, there are all sorts of varieties out there, perhaps there is a case to be made to take a wait and see approach on this.

Mr. CANNON. Thank you.

Mr. Cordi?

Mr. CORDI. Yes, let me start out by agreeing with your general observation. Obviously, flat taxation on basic services or goods that the whole population buys tend to be regressive. You are right. What we are looking at here, though, interestingly would have—this preemption would have the opposite effect, because, of course, who would avoid taxation as a result of this is necessarily people who are computer-literate, able to afford DSL connections, more sophisticated people; basically, the better off would be who would get the tax benefit, indirectly, frankly of this preemption, the immediate benefit of it.

And who is left holding the bag are all the people who will be locked into landline services for all the reasons that they cannot get this. So I would argue this preemption makes existing telephone taxes, as regressive as they may be, worse.

Mr. CANNON. Well, let me just follow up a little bit, because my sense is that people who have landlines in cities will tend to be close to DSLams or switches so they can get DSL relatively cheaply. It is the rural folks that have a long distance and are going to have a hard time getting DSL services that are left in a sort of a box. But I think as Mr. Pepper just said, those people in many cases are already getting broadband services.

So people who are living closely together and have landlines now are the people that are most likely to benefit from the plummeting costs of DSL, broadband or VoIP. It seems to me that—are we seeing the same issue, or am I missing something here?

Mr. CORDI. Well, that is correct as far as you are going. I guess my concern, though, is for those people who cannot take advantage of that, which is a very large chunk of the population that cannot throw up the money for a computer, get the cost of DSL, and those are the people who will be left using landlines, and frankly, my sense is preemption leaves this more regressive and not less regressive. That is an opinion.

Mr. CANNON. I do not mean to be tenacious about this, but you are going to have VoIP with just a phone. In other words, you will not even need a computer to do it. So you are not at the $1,000 or $500, I mean, today with Linux, you are probably at less than that for a computer that would work; in fact, we were pricing for Haiti refurbished computers at $100 a piece.

So the cost of a computer, I do not think is going to be a hurdle, and yet, the poorest tend to be the most closely-packed. They tend to have telephones already, and those are the folks who are going to lose a third to half of their phone bill by doing a VoIP, and a big chunk of that is tax, I grant you, but some of that is going to be purely economic, and over time, more of it will be purely economic. Are those not the very people that you want to bring into the—you want to not put on the wrong side of the digital divide?

Mr. CORDI. I will agree with that.

Mr. CANNON. Thank you, because here, we are not just talking about the taxation as being regressive. We are talking about the context being regressive. And I appreciate that information.

Let me ask all the witnesses about what happened in the Senate Commerce Committee yesterday. We passed, or they passed, Senator Sununu's VoIP bill that would preempt certain State taxes and regulations for 3 years but only 3 years. What potential problems do you see from a lack of certainty that is inherent in just a 3-year moratorium, or is the 3 years too much, whatever your view on that may be?

And Mr. Pepper, could we start with you and then just move through the panel?

Mr. PEPPER. We actually have not—I have not seen the latest language, but my understanding is that the 3-year moratorium language was a start in order to build a consensus to provide time for a more lasting approach.

Mr. CANNON. And so, you think 3 years is appropriate?

Mr. PEPPER. You know, I think we need clarity, you know. The question is, you know, will a 3-year approach at least provide clarity for 3 years while Congress considers what to do beyond that? I mean, that is my understanding from reading the trade press this morning. And clarity is the thing that investors need if they are going to roll out new services and make investments.

Mr. CANNON. Does 3 years provide enough certainty for investment, or is that too short a period of time?

Mr. PEPPER. I would ask the companies that question.

Mr. CANNON. That is a good point.

Mr. LANGHAUSER. We believe it should be permanent. We believe it should include VoIP. Certainty is just vital in this industry, and it is particularly acute to my company after what we have been through based on a flip-flop in Federal policy.

Mr. CANNON. Do you have, at the top of your mind, by any chance, the amount of money, the amount of capitalization that was lost from the top of the bubble to the bottom for telecom companies? My sense is something like $500 billion or $600 billion.

Mr. LANGHAUSER. I do not have that number in my head, but that sounds like a reasonable estimate.

Mr. CANNON. There have been a huge number which argues for clarity now and certainty now.

Mr. LANGHAUSER. Absolutely.

Mr. CANNON. Mr. Cordi?

Mr. CORDI. My understanding, and I am getting this only from press reports of what the Senate did, was they took out the State tax preemption language. The 3-year moratorium pertains only to regulation, and the tax language has gone away, but that is only from press reports, Congressman.

Mr. CANNON. Thank you.

Mr. KIRKLAND. We would support certainty again. We would echo the constant changes in the environment make it very hard to make investment decisions. And so, to the extent there can be a resolution that is at least permanent, obviously, nothing is permanent at the end of the day, but if—we prefer greater definition.

Mr. CANNON. Thank you.

Mr. Pepper, how long did it take the FCC to rule on the Pulver order, and how long did it take to rule on the AT&T order?

Mr. PEPPER. I would have to go back and check specifically. But I think it was Pulver was probably a little over a year, and I think the AT&T also—you may have the exact dates. About 18 months, probably about 18 months for each.

Mr. CANNON. Is there something that you can commit to for the FCC today about making these time frames shorter?

Mr. PEPPER. I wish I could make commitments on behalf of my bosses, but that is tough. We are working very quickly. I mean, literally, even here in July on a Friday, I talked to them this morning. We have staff working through the reply comments that came in on the notice of proposed rulemaking, so we actually are working on it, and we expect to have some pieces of this staff recommendations to the Commissioners by the end of the year.

Mr. CANNON. Six months there, 18 months there; the shelf life of these technological products is relatively short. We would encourage you that.

Mr. Pepper, in 2004, the FCC issued its order declaring AT&T's—yes, this is actually different, AT&T's IP telephone service was not exempt from paying the access charges applicable to circuit-switched interexchange calls. At a hearing a few weeks ago before the Energy and Commerce Committee, FCC Senior Deputy Chief Jeffrey Carlisle stated that the order applies only to AT&T until the broader VoIP questions are addressed in the IP-enabled services NPRM.

However, we have received information that despite the narrow read of the FCC order, incumbent carriers have applied this decision to other VoIP providers that are distinguishable from AT&T such as Calypso.com. We understand the incumbents continue to impose or threaten to impose access charges on these companies by misapplying the FCC order, what appears to be a misapplication of the FCC order.

In essence, the incumbents are freezing out the VoIP providers either directly or through threats to competitive carriers. What should be done about companies such as Calypso.com whose viability is threatened by a distorted interpretation of the AT&T order?

Mr. PEPPER. We became—first of all, the AT&T decision applies only to the specific facts of the AT&T case, and so that is absolutely correct. And we have learned over the last week or so of these kinds of actions on the part of incumbent carriers wanting to impose access charges on other forms of voice-over-IP on which the Commission has not yet made a determination.

So my recommendation to Calypso is to come in and talk to the people at the Commission. We also have other petitions pending as well as the notice of proposed rulemaking that is addressing situations that go beyond the AT&T set of facts.

Mr. CANNON. Thank you. Could you tell us a little bit about the FCC's efforts to address the social issues associated with VoIP, including universal service and 911 service?

Mr. PEPPER. Yes; we believe that it is very important that we separate economic regulation from what we call the social or consumer policies. Those include things like affordable phone service, access for law enforcement for first responders, access for people with disabilities, access for lawful intercept. And the Commission began a series of what we call solution summits, bringing the parties together to work through these issues. And for example, we had a solution summit with the law enforcement community and service providers to focus on 911 issues for first responders.

And frankly, there has been a lot of progress. There was wide agreement, for example, in that particular meeting that, number one, the voice-over-IP providers who were there said, you know, they actually believe it is important as a competitiveness necessity to provide 911 service going forward. There are some technical difficulties on figuring out location-based for certain forms of voice-over-IP, and they are working with the National Emergency Numbering Authority, NENA, which is the body which does the work for the law enforcement community and first responders, hospitals, fires and so on, firefighters, to work through those issues.

And in fact, on December 1, the two communities entered into a memorandum of understanding for short-term agreements while they work through long-term solutions. And I have a copy of that here if you would like to have that in the record.

Mr. CANNON. I would appreciate that for the record.

I mentioned earlier that I had been to the AT&T presentation. They have a registration process which allows you to put an address in, and from what I understand from what you're saying is there is now a context for that address to be useful, and I suspect in most cases, it would be useful to a local emergency responder.

Mr. PEPPER. Well, this is what they're working through. One of the questions for the first responders and the public safety access points is whether or not those what we call PSAPs actually have the equipment that could do something with that information. And so, part of the answer is funding for and upgrading the local first responder facilities, not just doing something with the voice-over-IP technology on the service provider side.

This, by the way, is very analogous to the issues with having location-based e-911 for mobile wireless, cell phone service, right, where the industry is, you know, stepping forward and providing it on their networks, but there are many of the local authorities that have not yet upgraded their facilities, because they just do not have the funds to do that.

Mr. CANNON. Thank you.

What efforts has the FCC made to address the issue raised by Mr. Langhauser concerning intercarrier compensation with regard to VoIP, and what is the position of the FCC on this point?

Mr. PEPPER. Well, the Commission and individual Commissioners have said that resolving the intercarrier compensation questions are among our highest priorities. And the reason is very simple: the intercarrier compensation arrangements that have grown up over the last 40, 50, 60 years were based upon monopolies and a single form of communications.

And essentially, what has happened is that we now have many competitive providers, and we have different prices for the same thing. What I mean by that is that we talk about intercarrier compensation; essentially what we are talking about is what one provider of service pays another to terminate a call. Those prices, and by the way, if you are the local carrier, the cost of terminating a call from your central office to your home or office is the same no matter where that call originates from.

Today, we have a regime in which if the call originates across the country, you pay one price; if it originates within the State but not your community, you pay another price; if it is from across town, you pay—a carrier pays a third price. If it's a cell phone company, you pay a different price. There are multiple prices for the same thing, and as a result, there is significant incentive for arbitrage.

And to give you an idea of the range of prices for this termination, if you are AT&T, and you are providing a long-distance call, and you want to terminate it, and it comes across the country, and you are going to a major, a big Bell company, you will pay about a half a cent per minute to terminate that. On the other hand, if you are taking that call to a small telephone company, the rural telephone companies, and the call originates within the State, for instance, Wisconsin, there is a small rural phone company in Wisconsin that has an intrastate access charge, in other words, intrastate termination charge of 12 cents per minute.

That is not sustainable going forward, because everybody eventually is going to have services like their wireless phone, where your local calling area, in terms of your pricing, is the United States. So we think this is extremely important. We have an open proceeding. There are industry negotiations, and this is one of the things that we are going to be working toward as soon as we can.

Mr. CANNON. Is this an issue that the FCC expects to resolve in its notice of proposed rulemaking on IP-enabled services?

Mr. PEPPER. No, there is a separate proceeding on intercarrier compensation.

Mr. CANNON. Thank you very much.

This has been an extraordinarily informative hearing. I appreciate the depth of understanding and clarity of statements. Are there any things that any of you would like to add at this point to the record?

It has also been, given the contentious nature, the difficult nature of it, it has been remarkably agreeable. I think that we understand—and, in fact, if I can just comment on the course of this, 3 months ago, I had people telling me that the 911, it was never going to work, and that was a terrific difficulty. We have made dramatic progress in recent times, and I think if I can characterize this hearing, there is dramatic consensus on the nature of the transition but concern about how we deal with that transition, especially from the point of view of the States and State revenues, because this is a larger threat, I think you have indicated, than the SSTP may represent, and so, we have to—let me just say that it is going to be extraordinarily important that we grapple with this.

It is just not acceptable to have the Senate stop stuff because one Senator can put a hold over there, because stopping is not going to change the course. And stopping may just end up leaving the States in much worse condition than if we are thoughtful and work out a process for resolving it. So, Mr. Cordi, I really appreciate your insights, the clarity of your thinking. I understand the urgency of it. And I am committed to helping, at least from this Committee's perspective, helping VoIP move forward, because I think it solves a host of problems, including for the poorest among us, recognizing that if that happens, something else has to happen to create a balance.

And so, I appreciate your input, especially, Mr. Cordi. I think it has been very thoughtful, very helpful and very agreeable, and I appreciate the technical and other kinds of input that we have gotten from the other panelists, which have been most enlightening.

Thank you, and this hearing is adjourned.

[Whereupon, at 11:30 a.m., the Subcommittee was adjourned.]

APPENDIX

Material Submitted for the Hearing Record

The VON Coalition

9-1-1 EMERGENCY NENA

NENA and Internet communications providers have agreed upon the following action items:

1—For service to customers using phones that have the functionality and appearance of conventional telephones, provide 9-1-1 emergency services access (at least routing to a PSAP 10-digit number) within a reasonable time (three to six months) and prior to that time inform customers of the lack of such access.

2—When a communications provider begins selling in a particular area, it should discuss with the local PSAPs or their coordinator (as identified on the NENA website) the approach to providing access. (For example, if routing to 10-digit number, confirm the correct number with the PSAP.) This obligation does not apply to any "roaming" by customers.

3—Support for current NENA and industry work towards an interim solution that includes (a) delivery of 9-1-1 call through the existing 9-1-1 network, (b) providing callback number to PSAP, and (c) possibly in some cases, initial location information. The current timeline for the NENA VoIP/Packet Committee to develop its interim recommended solution is May 2004.

4—Support for current NENA and industry work towards long-term solutions that include (a) delivery of 9-1-1 call to the proper PSAP, (b) providing callback number/recontact information to the PSAP, (c) providing location of caller; and (d) PSAPs having direct IP connectivity. The initial standards development work of the NENA VoIP/Packet Committee should be completed by the end of 2004.

5—Support for an administrative approach to maintaining funding of 9-1-1 resources at a level equivalent to those generated by current or evolving funding processes.

6—Consumer education. This could include projects involving various industry participants and NENA public education committee members to create suggested materials explaining any 9-1-1 differences to customers.

Voice on the Net Coalition

The voice for Internet voice innovation and the policy framework that enables it

UNLEASHING THE FULL PROMISE AND POTENTIAL OF INTERNET VOICE COMMUNICATION

Vast benefits: lower prices, better jobs, and improved ways to communicate

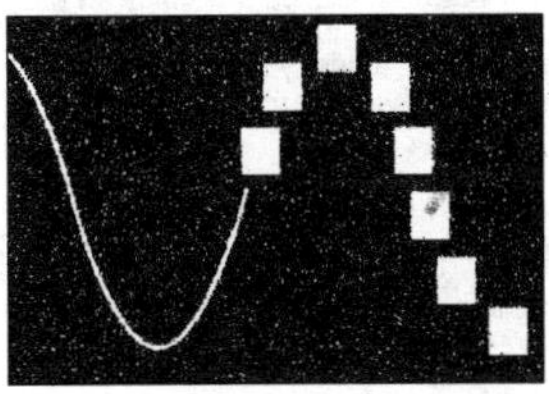

UNLEASHING THE PROMISE OF INTERNET VOICE COMMUNICATION

ACHIEVING VAST NEW BENEFITS WITHOUT THE NEED FOR LEGACY PHONE REGULATION

The Voice on the Net (VON) Coalition

Over the last two decades, Internet innovations like e-mail, the world-wide-web, and e-commerce have unleashed powerful transformations that that have changed almost every aspect of our lives, grown our economy, and increased our standard of living. On the horizon are a new wave of Internet based voice advances that promise to make talking more affordable, businesses more productive, jobs more plentiful, and the Internet more valuable.

Innovators around the globe are pumping out promising new voice applications that use high-speed Internet to deliver old services in fundamentally new ways. They are on the road to a dramatic transformation in the way we communicate – converging Internet text, voice and video in entirely new ways. At the forefront of this revolution is a new generation of Voice over the Internet (VoIP) advancements that harness the power of the Internet to transform the way we can talk.

To help promote this future, the nation's leading VoIP companies, on the cutting edge of developing and delivering voice innovations, have come together to advance regulatory policies that enable Americans to enjoy the full promise and potential of VoIP. The group of companies who make up the Voice on the Net or VON Coalition, believe regulators should refrain from applying traditional telephone regulations that could stall consumer benefits, while industry and government find new solutions and ways to address important concerns without imposing legacy telephone regulations to VoIP.

The VON Coalition includes AT&T, BMX, Callipso, CallSmart, Convedia, Covad, IceNet, iBasis, Intel, Intrado, MCI, Microsoft, PointOne, Pulver.com, Skype, TeleGlobe, Texas Instruments, VocalData, and Voiceglo. Together they believe that American's are fundamentally better off with a generally hands off regulatory approach to Internet and Internet based services like VoIP. Since its inception, the VON Coalition has consistently advocated that federal and state regulators maintain current policies of refraining from extending legacy regulations to Internet services, including VoIP. More information about the VON Coalition can be obtained at the following website: http://www.von.org.

The Coalition understands that VoIP is and can continue to deliver important benefits including:

- dramatic cost savings for consumers
- reduced operational costs for providers
- innovative new features for users
- increased competition for communities
- greater infrastructure investment
- accelerated broadband deployment
- improvements in emergency services
- lower cost communications for rural America and government users
- increased access for persons with disabilities, and
- increased productivity for our economy.

The potential for a vast new wave of VoIP-led technological innovation is here. But in order to unlock the vast new productivity, economic growth, and consumer benefits that lie ahead, policymakers need to help overcome a set of emerging policy challenges and nurture future innovation. To achieve these many benefits, the FCC, Congress, and the states need only maintain their successful hands off approach to regulating all forms of Internet communications.

The VON Coalition asks policymakers to classify VoIP as unregulated information services subject exclusively to federal jurisdiction. The Coalition acknowledges that there are important social policy issues where the FCC and state regulators have a legitimate role. But the companies believe these issues can be more effectively addressed without imposing heavy handed legacy telephone regulations to innovative Internet voice communications. For example, the VON Coalition supports efforts to address critical issues like the availability of 911 emergency services and disability access through voluntary and other efforts that that don't require imposing heavy regulation that could stifle voice innovations.

VoIP innovation can be:
- ✓ a force for increased competition
- ✓ a platform for innovation
- ✓ a driver of broadband deployment
- ✓ a vehicle for consumer benefits, and
- ✓ an enabler of economic growth

With respect to Universal Service, the Coalition favors reforming the Universal Service Fund to ensure its sustainability through a system of fair contributions from all providers through either a telephone number-based or connection based contribution system. Likewise, as a way to ensure fair compensation for carriers, the group favors an overhaul of the outmoded inter-carrier compensation regime which is now a hodgepodge of implicit subsidies.

This paper previews a vision for a new communications future and the Internet voice technologies that can drive it. It provides key insights into the breakthrough voice applications that will unleash the next wave of Internet innovation -- its benefits and challenges – and outlines the concrete steps that policymakers must take to realize its potential.

How VoIP Transforms Internet Communications

The technology behind VoIP works very much like e-mail. Rather than connecting directly over a single dedicated open circuit, the data is sent over the Internet in packets of data and reassembled on the receiving end. Because voice data packets can be interspersed between other e-mail and web page traffic on the Internet, the process doesn't use as much bandwidth and makes phone calls essentially as cheap to transmit as e-mail.

The combination of "IP" and "voice" completely changes the nature of voice from a simple utility service into a multifaceted information application just like e-mail, instant messaging, and video conferencing. The Internet changes voice communications into a software application allowing voice to be programmed, transformed, and converged in entirely new ways. It also allows voice communications to be integrated into almost any type of device, application or service that uses a microprocessor or touches the Internet.

Sending voice over IP networks makes all of this device diversity possible. IP blurs the functional distinctions between applications and devices: a computer can become a phone, a Wi-Fi handset can become a global intercom; web pages can become voice portals; and instant messenger software can become voice and video chat applications. As with other Internet traffic, IP renders distance irrelevant for voice traffic and makes phone numbers location independent.

VoIP is Not Traditional Telephony

VoIP is not another flavor of telephone service. It's a new frontier in communications for individuals and businesses alike. Until now, the limitations of technology have tied voice directly to a specific kind of physical network. The voice service and the physical infrastructure were one and the same. But the IP network decouples Voice from the copper telephone network, making it also potentially available on cable, fixed wireless, fiber, satellite and any other place IP is available.

VoIP is Delivering Real Benefits:

- ✓ Speeding new innovations to consumers and businesses
- ✓ Giving Americans more communications choices than ever before
- ✓ Driving vigorous new investment in the network and restoring vitality to the telecom sector

Voice is simply another application being deployed on the Internet, often in combination with other applications. These applications are possible, in part, because the Internet offers openness, thereby encouraging innovation. In contrast, the PSTN operates as a closed system on which it is impossible for innovative developers to build new applications. The failure of Advanced Intelligent Networking illustrates the problem of closed systems impeding the development of innovative products and services. The Internet permits entrepreneurial firms to develop new hardware and software applications that can seamlessly fit into the network. As computer processing power increases, IP-enabled products and services are poised to make communications more innovative, affordable, and universal.

People are adopting VoIP not just because it offer enormous consumer savings, but because it also provides innovative new features such as the ability to access voicemail from your e-mail, to conference large groups of people together, to select which area code your telephone will use, or to use your phone extension anywhere there is an Internet connection. In fact, once converted into IP, voice can be integrated with any number of software and data applications. This means rich new possibilities for innovators, businesses and consumers alike.

One new kind of VoIP service allows anyone with a broadband connection to plug their phone into a broadband router and make calls over the Internet. Entrepreneurs, such as Vonage, IceNet, PointOne, Voiceglo, and even AT&T have begun offering innovative voice applications to residential and small business consumers who have broadband connections including unlimited local and long-distance calling and on-line call logs. Vonage, for example, recently announced a package of unlimited local and long distance calling for $29.99 per month.[1]

Some innovative VoIP services originate and terminate on the traditional telephone network, but are only possible though use of an advanced IP communications network and are not possible or practical with use of only the legacy circuit-switched

[1] Vonage's service offerings are available at http://www.vonage.com/rate.php

network. For instance, PointOne's Star (*) Call IP service provides users with access to real-time information such as stock quotes and driving directions, the ability to communicate with IM, as well as other advanced features. The service is available during every voice communications session on the PointOne network. Users access this information by "dialing" a predefined key combination at any point during a "call" during which time the other user is placed on hold. When the user is done accessing this feature, the call is rejoined.

With Free World Dialup ("FWD") 3.0 or Skype software, users of different broadband technologies (cable, DSL, Ethernet, satellite, etc.) can place calls over the Internet to other users of the same software without ever accessing the PSTN. Unlike a traditional calling arrangement in which long distance calls generate usage-sensitive charges, FWD and Skype subscribers use a broadband connection and VoIP capability to make calls almost for nothing[2].

Advantages of IP Voice over Traditional Phone Circuits
The Internet and other IP networks offer an inherent efficiency, reliability, and functionality for communications, particularly those that combine different kinds of data, including digital voice traffic. The conventional circuit-switched phone network or PSTN works on the model that each customer's equipment must have a continuous connection (a "circuit") to a telephone company switch, whether or not the connection is actually in use. For long-distance services, a continuous circuit must be established and maintained between each pair of users for the duration of a call, regardless of the amount of information sent through that path. By contrast, the Internet trades increased use of computer processing for a decreased use of transmission facilities and automatically re-routes packets around problems such as malfunctioning routers or damaged lines, without relying on a separate signaling network. As the cost of computer processing continues to decrease and the demand for communications bandwidth by consumers increases, IP networks increasingly offer a more economical and robust means for providing communication connections.[3] Moreover, unlike the PSTN, where service providers must either build their own or rely on the incumbents' infrastructure, the Internet allows new competitors to swiftly emerge because they do not need to own or construct any infrastructure.

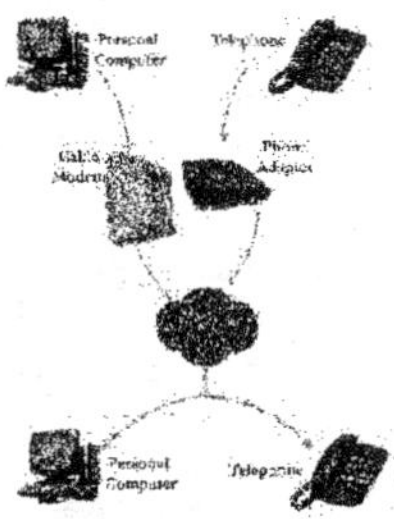

Source: FCC

Impact and Benefits of VoIP

The development of IP-enabled services, including VoIP, is having a profound and beneficial impact on the United States and the world. Below are just some of the benefits of IP-enabled services.

---

[2] Similar to the cost of e-mail, consumers must pay for their own broadband or other transmission to make Skype or FWD calls.
[3] *Petition for Declaratory Ruling that AT&T's Phone-to-Phone IP Telephony Services are Exempt from Access Charges, Order*, WC Docket No. 02-361, FCC 04-97 (April 21, 2004) ("*AT&T Declaratory Ruling*"), at ¶ 3 ("VoIP uses available bandwidth more efficiently than circuit-switched telephony and allows providers to maintain a single IP network for both voice and data.")

VoIP can deliver dramatic consumer savings

Because VoIP converts voice into Internet data and routes the data as packets like e-mail, a voice call takes up less transmission capacity – using up to 90% less bandwidth than a traditional PSTN call -- making VoIP calls less expensive and more efficient[4]. Taking advantage of these differences has enabled companies to offer phone-like services that cost as little as $29 a month for unlimited local and long distance calling. Experts estimate Internet telephony could save consumers between 40 percent and 60 percent on their phone bills.[5] Another study found that the average narrowband household could capture a net savings of $8 per month if they upgraded to broadband and began using a VoIP telephone service.[6]

In the United States, hundreds of thousands of immigrants use VoIP to dramatically lower the cost of communicating with friends and relatives outside of the United States, through either personal computer-based VoIP or VoIP used by prepaid calling card companies. Phone-to-gateway network configurations, such as those offered by Callipso, provide those without a computer or broadband service what is often their only access to the benefits of the Internet.

If international voice service is any guide, VoIP stands to spark new competition and further drive down consumer prices. Rates for international calls have dropped 80% over the last two decades. Much of that decline stems from cheap VoIP service carrying up to 12% of international calls[7].

VoIP is cheaper to deploy too.

VoIP networks are often based on software rather than hardware, which is easier to alter and maintain, and helps reduce operation costs. Some experts claim that installing a packet-switching network costs about a third of a circuit switching system and that one can save about 50-60% in operating costs.[8] The potential cost saving through converging voice and data applications in one network makes VoIP attractive for enterprises that have already deployed an IP network. Others report the savings are even greater. The Precursor group reports that a SIP softswitch is about *one-tenth* the cost of a circuit switch, on a one-for-one replacement basis[9]. With scale, they report it is roughly *one-thirtieth* the cost.

VoIP provides innovative new features

People are adopting VoIP not just because it offer enormous consumer savings, but because it also provides innovative new features such as the ability to access voicemail from your e-mail, to conference large groups of people together, to select which area code your telephone will use, or to use your phone extension anywhere there is an Internet connection.

---

[4] http://www.fwcs.co.uk/voip.htm

[5] The Detroit *News Internet phone use grows: Less costly service is to be offered by major firms in '04* By Charles E. Ramirez , December 28, 2003

[6] Market research firm Parks Associates, study February 2004

[7] Internet Calling Posing A Threat To Landline Phone Companies, By Mike Angell Investors Business Daily 5/18/04 http://www.investors.com/editorial/general.asp?v=5/19

[8] Equipment manufacturer Sonus Networks estimates that installing a packet-switching network could be done for about a third the cost of a circuit switching system, and that operating savings could be 50 percent to 60 percent. New York Times, January 12 2004, http://www.nytimes.com/2004/01/12/technology/12phone.html?th=&pagewanted=print&position=

[9] Precursor Group, SIP "De-geograph-ies" Telecom: Transforms Central Office Assets into Liabilities, May 5, 2004

As Newsweek points out, using VoIP, "clever Web interfaces will let you convert your voicemail messages to email, or your emails to voice; you'll be able to call-forward in a myriad of ways, or switch to video or hi-fi voice if you want, or even agree to hear some number of commercials every day to lower your bill.[10]"

VoIP is now being integrated into a variety of new applications like[11]:

- A PDA that uses Wi-fi for voice
- A Webcam using VoIP for videoconferencing
- Instant messenger software converging voice, text and video chat
- A Wi-fi enabled badge that allows wearers to touch the device, say who they want to talk to and be connected – much like Star Trek.
- Game consoles that allow gamers with a headset to talk with each other during team-based games

Consumers are getting innovative new features, including:[12]

- **Web-based call logging** - Users check personal Web pages for detailed, up-to-the-minute listings of incoming and outgoing calls.

- **"Find me, follow me"** - Users program their phones to search for them when a call is not answered. A work phone, for example, would ring over to a cellphone if not answered. If that, too, went unanswered, the call might be forwarded again to a home phone or other number.

- **Do Not Disturb** - Instead of permitting the phone to ring at all hours, incoming calls can be sent directly to voice mail at scheduled times, eliminating calls during dinner, during a favorite TV show or after bedtime. (A work-around would still allow urgent calls to ring through.)

- **Digital voice mail** - Voice mail messages are converted to digital sound files that can be easily stored or e-mailed. Users might have voice mails from their home phone forwarded by e-mail to their desk at work.

- **A truly cordless phone** - VOIP phones can be plugged into any broadband connection, so users can take their home phone on vacation, on business trips, or while visiting friends and family.

- **Customized voice mail boxes** - Incoming calls can be routed to special voice mail boxes based on caller ID. Business clients might get one voice mail greeting; family members another.

Consumers are excited about VoIP possibilities. A recent survey found that three-quarters of adults have heard of VoIP – 4 times more than have heard of Wi-fi.[13] Approximately 2 of 3 believe VoIP will forever change how we communicate.

---

[10] Newsweek, *Will Telephone calls be free?*, May 18, 2004 http://msnbc.msn.com/id/3730179/site/newsweek/
[11] Werbach testimony to FCC at VoIP hearing http://werbach.com/docs/FCC_VOIP.txt
[12] Cyberphones are no longer just for nerds, by John M. Moran -- Hartford Courant 05/31/2004
[13] According to an Ipsos-Insight Express study, commissioned and released by AT&T March 2, 2004, this was most evident in results that found 74 percent of the sampled population of 1,000 adults has heard of some form of VoIP. That percentage is approximately 4x the consumer awareness of Wi-Fi (19 percent), a popular wireless high-speed connection technology, and exceeds that of DSL (66 percent), a high-speed

In fact a majority of consumers believe VoIP's impact will be similar to digital music, flat screen TVs, and computer games. Further evidence suggests that consumers believe the technology will emerge rapidly. Nearly one in two consumers believe phone service will move to the Internet within two years (47 percent) and nearly one in four (23 percent) of those respondents believe it will happen this year. And even non-users are interested. Among current "non users" aware of VoIP services, 76 percent would consider actually implementing the service in the next year, depending on the price and package offering.

VoIP Is Enabling Competition in New Kinds of Voice Services
As a result of VoIP technological advances, VoIP providers and start-ups are charging into the Internet voice market at unprecedented rates. In fact, almost every major telecommunications provider now has plans to introduce Internet-based voice service to take advantage of the technology's lower costs, its capacity to deliver new innovative services, and ability to compete in the local phone market.

This may be just the beginning. Tomorrow almost anyone with a bright idea and access to the Internet can jump into the game. VoIP is enabling a host of new non-traditional competitors to enter the voice market which was once the sole purview of a incumbent telephone. Software, hardware, networking gear companies and innovators in their basements are now able to program new voice applications and become global voice providers with the reach of the Internet.

For consumers, the benefits of this independence are profound. With a broadband connection, consumers will be able to choose directly the type of services and the specific provider they want to deliver it independent of the incumbent broadband provider, and regardless of the type of services the incumbent has to offer.

Internet voice is calling:
- ✓ 2/3rds of Americans believe VoIP will forever change how we communicate
- ✓ Consumer achieving savings of 40 to 60 percent off phone bills
- ✓ VoIP equipment investment increasing as much as 50% in 2004
- ✓ 30% of U.S. businesses plan to move to VoIP within the next two years

VoIP Is Accelerating Investment in Innovation Infrastructure
This Internet induced competition is real and the adoption curve is arching steeply skyward and breathing new life into the tech sector – helping increase investment in the Internet, the high-tech sector, network and service providers.

U.S. carriers spent an estimated $2 billion on VoIP equipment in 2003, an increase of approximately 10 percent from 2002.[14] In contrast, spending in the overall telecommunications equipment market declined by 20 percent in 2003. Investment in VoIP equipment is estimated to grow by 50 percent in 2004.[15] As the

---

method to move data over phone lines. Nearly two thirds of the participating consumers (63 percent) would go so far as to say, 'VoIP will change the way they communicate.';66 percent suggested VoIP's impact on their lives will be similar to that of digital music; 58 percent compare it to computer games, and 57 percent a flat screen TV. www.att.com/presskit/voip.

[14] Steve Rosenbush, *Telecommunications: Strong Signals the Bad Times Are Over*, Business Week, January 12, 2004, at 100.

[15] Similarly, one study predicts that the market for all VoIP equipment, about $1 billion in 2002, is likely to reach almost $4.3 billion in 2006. *See Getting the Value From VOIP* (November 13, 2003) (available at:

telecommunications sector begins its recovery, VoIP will be essential to sustaining robust growth and investment.

VoIP Is Increasing Broadband Penetration. VoIP may be the long awaited "killer application" for driving broadband subscribership. There are already signs that consumers are flocking to broadband in order to take advantage of new broadband VoIP calling plans. The Yankee Group predicts that VoIP could spur new growth in untapped markets and enable entirely new business models. At present, only about 20 percent of Americans have subscribed to broadband.[16] Among those who do not have broadband, approximately 70 percent report that broadband is too expensive.[17] VoIP, however, can overcome price barriers by dispersing the cost across both products – voice and broadband. While broadband penetration rates currently drive VoIP adoption, VoIP could become the application to drive future broadband adoption.[18] One study estimates that widespread adoption of broadband could add $500 billion to the economy[19] and generate more than 1.2 million jobs.[20] In the next five years, the proliferation of VoIP services will create huge opportunities for consumers and even greater growth for broadband providers.

> *"What differentiates this period from other periods in our history is the extraordinary role played by information and communication technologies. The effect of these technologies could rival and arguably even surpass the impact the telegraph had prior to, and just after, the Civil War."*
> *– Alan Greenspan, July 13, 2000*

VoIP Can Increase Productivity And Help Lift the Economy While VoIP is driving new capital investment in both the Internet backbone and last mile, the biggest lift to the economy may come from major new productivity gains for companies. Productivity gains can in turn help lift the economy and overall standards of living. For example companies can become more efficient by deploying VoIP to help gain the benefits of telework and a distributed workforce.

- Dialing into a corporate IP PBX system can give home based workers and road warriors all of the features and benefits of the corporate VoIP network.
- By eliminating the commute, it provides workers with more hours and options for working.
- An IP PBX maintained at an emergency back-up site may mean the difference between business continuity and massive disruption during times of national or local emergency.
- Businesses are also finding they can connect remote offices and take a big bite out of the cost of keeping in touch with overseas divisions.
- VoIP also reduces operational costs because far fewer technicians are required to run and maintain a VoIP-based network.

---

http://www.lightreading.com/document.asp?doc_id=42831&site=lightreading) (citing *Infonetics Research Inc., Next Gen Voice Quarterly Worldwide Market Share and Forecasts* (August 2003)).

[15] The Yankee Group, *VoIP: Influencers and Drivers in the Emerging Broadband Telephony Market* (April 22, 2004).

[17] *Id.*

[18] *Id.*

[19] Robert W. Crandall and Charles L. Jackson, *The $500 Billion Opportunity: The Potential Economic Benefit of Widespread Diffusion of Broadband Internet Access* (July 2001).

[20] Stephen B. Pociask, TeleNomic Research LLC, *Building a Nationwide Broadband Network: Speeding Job Growth* (February 25, 2002) (available at: http://www.newmillenniumresearch.org/event-02-25-2002/jobspaper.pdf).

VoIP providers with the ability to cut corporate phone bills in half are finding a receptive audience among corporate financial officers.

This convergence of voice and data now allows companies to distribute work in new ways and literally eliminate the walls that once limited organizations. As an example JetBlue, the low cost airline start-up, has set up a "virtual call center" where 700 reservation agents work from home and answer VoIP calls that integrate passenger data with a consumer voice call. It cuts commutes, eliminates the need for a costly physical space, and drives major productivity improvements. A workers commute is as quick as a mouse click for the reservation agents, who use special application software on a VoIP Softphone. The software connects them to the airline's Internet telephony switch, which routes customer reservations calls to them. These work from home agents will handle 9.6 million phone calls in 2003. As a result, JetBlue's call-center attrition rate is just 5%, vs. 30% industrywide. That's helping JetBlue earn industry-leading profit margins of 19%. "They're a happier, more motivated, more loyal workforce," says CEO David G. Neeleman.[21]

Companies such as Dell and American Express currently are using VoIP in their call centers[22]. A recent report by In-Stat/MDR reported that at the end of 2002, 46.3 percent of enterprise phone stations sold in the United States were Internet protocol-based. It means that all businesses - small, medium and large - are tapping into VoIP to help decrease costs, increase efficiency and provide a better means of communicating. It is precisely the kind of technology that can drive the next generation of workplace productivity improvements.

VoIP Can Also Deliver Benefits to Specific Communities:

Emergency Services. VoIP is also able to deliver advanced emergency services, such as the ability to deliver reverse 911 and to conference in rescue workers on the way to an emergency scene.

> *"The 911 system is vital in our country, but it is limited functionally. In most systems, it primarily identifies the location from which the call was made. But an Internet voice system can do more. It can make it easier to pinpoint the specific location of the caller in a large building. It might also hail your doctor, and send a text or Instant Message alert to your spouse."*
>
> -- FCC Chairman Michael Powell

Some local governments are already using VoIP to deliver advanced emergency benefits. For example, Herndon, Virginia is using a VoIP system that automatically displays a picture of a missing child and possible suspects to VoIP phones equipped with special screens used by municipal workers.[23] Moreover, the Department of Commerce has combined its voice system with its emergency broadcast system, creating a reverse 911 system whereby users are contacted in the case of an emergency.[24] Eventually, IP networks will allow Public Safety Answering Points

---

[21] Business Week, 8-25-03

[22] Using the Web to make calls -- Quad-City Business Journal (IA) 05/31/2004

[23] *Net Phones Start Ringing Up Customers*, Business Week, December 29, 2003, at 45.

[24] William Jackson, "*With VoIP, Digital Department Comes of Age at Commerce*" (available at: https://secure.cio.noaa.gov/hpcc/docita/files/with_voip_digital_department_comes_of_age_at_commerce_09162003.pdf).

(PSAPs) to lower costs and to move more quickly in the event of an emergency.[25] Moreover, an IP-enabled PSAP will be better equipped to handle multimedia information and better respond to people with disabilities who may rely upon text- or video-enabled signing to communicate in an emergency.

Rural America. IP-enabled services are also benefiting rural America. One of the goals of universal service is to provide affordable voice communications to rural America, and no technology offers more promise for achieving this goal than VoIP.[26] Experts estimate Internet telephony could save consumers between 40 percent and 60 percent on their phone bills.[27] Moreover, VoIP is the application that will drive broadband deployment, including in rural America where access to broadband lags behind the rest of the nation.

Disabled Persons. IP-enabled services are also providing new opportunities for the disabled. The National Federation of the Blind uses IP-based phone services to provide a free newspaper reading service that uses voice synthesis to allow users to change voice speed and to search for words.[28] Avaya has just released a program that allows the functionality of a phone to be accessible to the blind without requiring any changes to the phone.[29] Blind employees at the Department of Education use IP communications to check e-mail remotely through the Department's voicemail system.[30] Trace Center and Gallaudet University are currently working with Cisco on a technique that would allow every phone within the organization to be instantly capable of text communication simply by installing a software program on the call manager server. This enables a deaf person to communicate in text (or in text and voice) without needing any special equipment and without changing the software on the phones.[31] The Washington School for the Deaf in Vancouver, Washington has used IP communications to afford equal access to communications services to deaf, hard-of-hearing, and hearing employees alike.[32]

Government. The federal government itself is adopting VoIP to achieve a wide variety of cost saving and service benefits. The Department of Commerce, Food and Drug Administration, Census Bureau,

People around the world are benefiting. Baghdad resident Usama Kamil Al-Sharqi is using using an Internet telephony feature built into instant messaging software to make phone calls from Baghdad Internet cafes that charge around $1 per hour for connect time, versus a going rate of about $1 per minute for long distance telephone service. (Washington Post)

---

[25] *Testimony of Professor Henning Schulzrinne, Department of Computer Science, Columbia University at FCC's Internet Policy Working Group E911 Solutions Summit* (March 18, 2004) (available at: http://www.fcc.gov/ipwg/E911SummitHenning.pps).
[26] *Testimony of Tom Evslin, CEO, ITXC, at FCC's VoIP Forum* (December 1, 2003) (available at: http://www.fcc.gov/voip/voipforum.html).
[27] Charles E. Ramirez, *Internet Phone Use Grows: Less Costly Service Is to be Offered by Major Firms in '04*, Detroit News, December 28, 2003.
[28] *Free Service to Those Who Cannot Read Regular Newsprint!* (available at: http://www.nfb.org/newsline1.htm); *see also* USA Datanet Corporation *ex-parte*, WC Docket No. 02-361 (February 2, 2004).
[29] Gregg C Vanderheiden Ph.D, *Access to Voice over Internet Protocol* (December 2003) (available at: http://www.tracecenter.org/docs/2003-NMRC-VoIP-Access/).
[30] News Release, *Cisco IP Communications System Improves Productivity for Disabled at Washington School for the Deaf and U.S. Department of Education (available at: http://newsroom.cisco.com/dlls/2004/prod_020904c.html).*
[31] Gregg C Vanderheiden Ph.D, *Access to Voice over Internet Protocol* (December 2003) (available at: http://www.tracecenter.org/docs/2003-NMRC-VoIP-Access/).
[32] News Release, *Cisco IP Communications System Improves Productivity for Disabled at Washington School for the Deaf and U.S. Department of Education (available at: http://newsroom.cisco.com/dlls/2004/prod_020904c.html).*

Environmental Protection Agency, Department of Defense, and Peace Corps, among other governmental entities, use some form of VoIP technology.[33] One study suggests that governments at all levels could save as much as $3-10 billion by using VoIP.[34]

International. Perhaps the most dramatic impact of IP-enabled services has been in certain foreign markets, where VoIP has been a leading force for lowering costs to consumers, increasing competition, and increasing deployment of broadband. According to Telegeography, international VoIP traffic increased by 80 percent to 18.7 billion minutes, and comprised approximately 10.8 percent of all international call traffic.[35] At the same time, rates for international calls have dropped 80% over the last two decades. Much of that decline stems from cheap VoIP service carrying up to 12% of international calls[36]. VON Coalition members have persuasively invoked the United States regulatory model in lobbying overseas governments, such that in former monopoly markets the first steps toward deregulation have included implementing low-cost VoIP.

Growth of VoIP and IP-Based Services

While IP-enabled services offer great promise, they are still in the nascent stages of development. The deployment of IP-enabled services, for example, has not had significant impact on the revenue of traditional, domestic, circuit-switched telephone companies. The use of VoIP by phone, in the enterprise setting, and by broadband consumers is not coming at the expense of phone company revenues. Moreover, IP-enabled services have not been demonstrated to have a significant impact on universal service or access charge revenues.

One factor contributing to this minimal impact is the current penetration rate for VoIP. While the number of Internet-based phone lines is projected to grow from well under a million in 2002 to more than 5 million by the end of 2004,[37] this represents a tiny fraction of the 113 million households where the traditional phone line will still be the primary line. Given that only approximately 60 percent of American households own PCs[38] and only 20 percent have access to broadband,[39] the number of people who can take full advantage of broadband-enabled VoIP applications is still limited.

---

[33] PlanetGov, *Multiservice/Convergence Technologies* (available at: http://www.planetgov.com/ns/consulting/solcontech.htm).

[34] *Government Could Save $3-10 Billion with VoIP, Study Says* (available at: http://www.forrelease.com/D20040211/nyw195.P1.02112004183814.23019.html) (citing study by Alexis de Tocqueville Institution).

[35] *NPRM* at n.34 (citing *Telegeography 2004*).

[36] Internet Calling Posing A Threat To Landline Phone Companies, By Mike Angell Investors Business Daily 5/18/04 http://www.investors.com/editorial/general.asp?v=5/19

[37] *Net Phones Start Ringing Up Customers*, Business Week, December 29, 2003, at 45 (citing study by Adventis Corp.).

[38] Jane Weaver, *Saying 'No Thanks' to the Internet: Online Growth in U.S. Flattens as Some Simply Opt Out* (April 16, 2004) (available at: http://msnbc.msn.com/id/3078958/); *NTIA, A Nation Online: How Americans are Expanding their Use of the Internet: a February 2002 Joint Study by the U.S. Economics and Statistics Administration and the National Telecommunications and Information Administration* (February 2002) (available at: http://www.ntia.doc.gov/ntiahome/dn/).

[39] Instate/MDR, *Reaching Critical Mass: The US Broadband Market* (March 2004) (available at www.instat.com).

"Hands-Off" Regulatory Approach to IP-Enabled Services

The growth of VoIP services has been propelled in part by the U.S. Government's "hands-off" regulatory approach. The hands off approach to Internet regulation has been an enormous success making the U.S. a leader in the development of VoIP and providing an influential policy model that has been emulated by many other countries. Since the inception of voice over the Internet, the FCC, states and Congress have consistently declined to regulate. The FCC articulated its policy in its 1998 *Universal Service Report to Congress*, which discusses various scenarios for what it called "IP telephony."[40] The *Report to Congress* discusses the difficulty of categorizing VoIP and the extent to which many of its deployments have characteristics of unregulated, information services.[41] As a result, the FCC expressly deferred any definitive pronouncements regarding VoIP, including phone-to-phone VoIP. As the FCC explained, "[w]e recognize that new Internet-based services are emerging, and that our application of statutory terms must take into account such technological developments. . . . We do not believe . . . that it is appropriate to make any definitive pronouncements [regarding VoIP] in the absence of a more complete record focused on individual service offerings."

> **Congressional "hands off" mandate:**
> 47 USC 230(b)(2)
> "(b) It is the policy of the United States
>
> (2) to preserve the vibrant and competitive free market that presently exists for the Internet ... unfettered by Federal and State regulation;"

Attempting to classify the dizzying array of IP-enabled services into statutory boxes is a Herculean task that would take arbitrary line-drawing that policymakers should avoid. As the FCC explained in a 1999 Working Paper, "[a]s more services are offered that use the Internet Protocol in a packet-switched environment, it becomes increasingly difficult to determine where the telecommunications service ends and the information service begins."[42] This statement is no less true today. IP-enabled services include a wide variety of network architectures, technologies, and applications. IP traffic travels as indistinguishable packets of digital bits, thereby blurring the lines between traditional services and categories.

An Inability to Force VoIP into Existing Regulatory boxes

The borderless nature of the Internet, and the inability to distinguish and regulate one bit differently from another, makes Internet communications a difficult thing to

---

[40] *Federal-State Joint Board on Universal Service*, Report to Congress, 13 FCC Rcd 11501, ¶¶ 83-93, 98 (1998) (*"Report to Congress"*) (also referred to as the "Stevens Report"). The *Report to Congress* addressed many of the issues raised in a 1996 petition for rulemaking asking that IP telephony software and hardware providers be classified as common carriers. *Id.* at ¶ 83 n.172; *see America's Carriers Telecommunications Association, Provision of Interstate and International Interexchange Telecommunications Service via the "Internet" by Non-Tariffed, Uncertified Entities, Petition for Declaratory Ruling, Special Relief, and Institution of a Rulemaking*, RM-8775 (filed March 4, 1996).

[41] As noted in a 1999 Commission Working Paper, "[a]s more services are offered that use the Internet Protocol in a packet-switched environment, it becomes increasingly difficult to determine where the telecommunications service ends and the information service begins." Jason Oxman, *The FCC and the Unregulation of the Internet*, OPP Working Paper No. 31, at 22. "Despite this difficulty, however, it remains important for the FCC to maintain the unregulated status of data services offered over telecommunications facilities." *Id.*

[42] Jason Oxman, *The FCC and the Unregulation of the Internet*, OPP Working Paper No. 31, at 22.

define, let alone regulate – especially in a technology that is constantly evolving. Ultimately, as networks move to an all IP-based world, all instant-messaging, video-conferencing, e-mail, IP television, and other technologies that utilize Internet communications are likely to have a VoIP component. The regulatory treatment decided today will have a dramatic impact on how these future technologies will emerge.

Exclusive Federal Jurisdiction Avoids 50 Different Regulatory Treatments
VoIP is interstate in nature and thus should be subject exclusively to federal jurisdiction. One of the inherent characteristics of IP-enabled services, and one of its advantages, is that it is entirely geographically neutral. There is no dedicated transmission facility required, there are no facilities required to be located locally. Internet traffic can travel anywhere in the world with no material difference in cost, and facilities which act on the call can be located anywhere.[43] Moreover, there is currently no method to identify or distinguish IP-voice from other IP traffic, or to determine the jurisdictional nature of the traffic. If VoIP were subject to state regulation, it would have to satisfy the requirements of more than 50 state and other jurisdictions with more than 50 different certification, tariffing and other regulatory obligations. Additional state regulation would eliminate any benefit of using the Internet to provide voice service. Certainly, this is kind of impact Congress considered when it made clear statements about leaving the Internet and interactive computer services free of unnecessary federal and state regulation.

Policy Goals Can Best Be Accomplished Through Voluntary And Other Efforts
There are important social policy issues where the FCC and state regulators have a legitimate role. But these issues can be more effectively addressed without imposing heavy handed legacy telephone regulations to innovative Internet voice communications.

Two examples:

- **911 Emergency Services.** The VON Coalition supports efforts to address critical issues like the availability of 911 emergency services through voluntary and other efforts that that don't require imposing heavy regulation that could stifle voice innovations. To advance these solutions, in December 2003 the VON Coalition joined with the National Emergency Number Association (NENA) to bring together leaders from the VoIP industry to forge a voluntary agreement on the next steps to develop the technical and operational mechanisms for providing effective access to emergency services by users of VoIP. NENA and members of the VoIP industry agreed on a set of key elements for providing E911 to VoIP users.[44]
- **Disability Access.** The IP-enabled services industry has also undertaken voluntary efforts to ensure that persons with disabilities are provided access to IP services. For example, the IP-enabled services industry has worked to develop and implement technology that is interoperable with TTY devices. The deployment of Internet

---

[43] *NPRM* ¶ 4 ("Packets routed across a global network with multiple access points defy jurisdictional boundaries.").

[44] *See Press Release, Public Safety and Internet Leaders Connect on 9-1-1* (December 1, 2003) (available at: http://www.nena.org/NENAVONVOIP%20press%20release%20FINAL%20112603.pdf).

enabled services has already had positive implications for access to communications by the hearing impaired. For instance, video relay service, an Internet-based video interpreting service for the deaf, now offers callers options involving web cameras for sign language.

Existing Economic Subsidy Mechanisms Must Be Reformed
Today's universal service system and inter-carrier compensation regime are in dire need of reform. The FCC needs to first reform these existing regulatory frameworks before considering whether and how they should apply to VoIP.

- ***Universal Service Reform.*** In today's economy, it is more important than ever that consumers in rural areas have access to affordable communications services and innovative new technologies regardless of where they live. The VON Coalition has long supported the goals of the universal service program. Even under the current USF regime, VoIP providers contribute to universal service either directly or indirectly. When an information service provider purchases an underlying telecommunications input, this generates indirect contributions to universal service support mechanisms.

  The VON Coalition believes that VoIP already meet the goals of universal service and need not be subject to these regulatory regimes. But before the Commission even considers whether to impose these requirements on VoIP, the FCC must first reform the existing regulatory frameworks. The VON Coalition is committed to bolstering and reforming universal service in a way that puts universal service on a more solid financial footing going forward and supports a numbers or connections based approach. A numbers- or connections-based contribution mechanism would better ensure the continued sustainability of the Universal Service Fund than any attempt simply to include IP-enabled and other information services in the current revenue-based mechanism.

- ***Inter-Carrier Compensation Reform.*** As for inter-carrier compensation, the Commission should move away from a hodgepodge of implicit subsidies and towards a rational series of voluntary inter-carrier business arrangements.[45] It's a broken system in need of reform. Applying a broken system to new technologies could stifle innovation and consumer benefits. Applying access charges on any class of VoIP service is unnecessary because incumbent phone companies are already fully compensated for their costs when Internet phone calls are terminated on their networks. When a phone company terminates VoIP services on its network, the phone company receives either reciprocal compensation and/or local end-user business line rates – in either case fully compensating them for the cost of the network. At the federal level, Congress has required the FCC to eliminate inefficient implicit subsidies from interstate access charges.[46] Rather than imposing legacy access charges adopted for a 100 year old telephone

---

[45] The impact of VoIP on access charges revenue is minimized by current rules governing access charges that accommodate ISP usage. Under an access charge exemption dating to the 1980's, ISPs compensate local exchange carriers through the purchase of business lines, not switched access.

[46] Chairman Powell recently remarked that "We must make all implicit subsidies explicit to ensure continued high-quality, affordable service and network investment. To that end, I applaud those states that have undertaken efforts to adjust retail rate structures and intra-state access charges." *Remarks of Michael K. Powell, Chairman, FCC, at the National Association of Regulatory Commissioners General Assembly*, Washington, DC (March 10, 2004) (available at: http://hraunfoss.fcc.gov/edocs_public/attachmatch/DOC-244737A1.doc).

network on innovative Internet communications services, the Commission should overhaul the current access charge regime.[47] "Bill and keep" may well turn out to be an effective arrangement as it has been in much of the IP world.

CONCLUSION:

VOIP is one of the most significant advancements in communications since the arrival of e-mail. The potential for a vast new wave of VoIP-led technological innovation is here. It has the potential for speeding investment, innovations and economic growth. It can deliver greater choices, lower costs, higher broadband demand, and innovative new services. In order to unlock these vast new benefits, policymakers need to help overcome a set of emerging policy challenges and nurture future innovation. Policymakers need only maintain their successful hands off approach to regulating the Internet for all forms of VoIP.

About the VON Coalition:
The VON Coalition consists of leading VoIP companies, on the cutting edge of developing and delivering voice innovations over Internet. The coalition, which includes AT&T, BMX, Callipso, CallSmart, Convedia, Covad, IceNet, iBasis, Intel, Intrado, MCI, Microsoft, PointOne, Pulver.com, Skype, TeleGlobe, Texas Instruments, VocalData, and Voiceglo, believe that American's are fundamentally better off with a generally hands off regulatory approach to Internet and Internet based services like VoIP. Since its inception, the VON Coalition has consistently advocated that federal and state regulators maintain current policies of refraining from extending legacy regulations to Internet services, including VoIP. More information about the VON Coalition can be obtained at the following website: http://www.von.org or by calling Jim Kohlenberger at (703) 237-2357.

[47] In 2001, the Commission initiated a proceeding to revise the intercarrier compensation regime. *Developing a Unified Intercarrier Compensation Regime, Notice of Proposed Rulemaking*, 16 FCC Rcd 9610 (2001).

VOICE ON THE NET (VON) COALITION POLICY PRINCIPLES

The longstanding U.S. policy of "hands off the Internet," has been emulated by governments everywhere and has been an enormous success. VoIP is a force for increased competition, a platform for innovation, and a driver of broadband deployment.

> *"I believe that IP-based services such as VOIP should evolve in a regulation-free zone. No regulator, either federal or state, should tread into this area without an absolutely compelling justification for doing so."*
> -- FCC Chairman, Michael Powell

The best public policy is to refrain from applying traditional telecom regulation to VoIP and to affirmatively create a national policy vision that ensures that traditional telecom regulation does not apply to Internet voice communications throughout the country.

While members of this coalition have different views on how much market power some facilities-based telecom providers have, we all agree that policies should be continued that permit entities that do not have significant market power to deploy voice over IP free from traditional telecom regulation.

The Coalition freely concedes that there are important social policy issues where the FCC and state regulators have a legitimate role. The VoIP community is prepared to work constructively on such issues, including providing access to those with disabilities, access to emergency services, cooperation with law enforcement, secure funding for universal service, and reform of inter-carrier compensation. These legitimate concerns can be addressed without imposing heavy regulation on VoIP and that if they are addressed successfully the pressure to regulate VoIP will dissipate. The coalition supports efforts to address these issues including:

- **Emergency Services.** VoIP industry representatives have been working with the National Emergency Number Association's ("NENA's") VoIP/Packet Technical Committee and VoIP Operations Committee to assess the current state of 911 provisioning in VoIP environments and to develop 911 solutions. There are important differences between the provision of 911 for traditional PSTN traffic and for VoIP, but there is every reason to expect that technical solutions exist to provide users with reliable access to public safety services. NENA and representatives of the VoIP industry recently reached a voluntary agreement on the next steps to develop the technical and operational mechanisms for providing effective access to emergency services by users of VoIP.

- **Law Enforcement.** Voluntary efforts also are underway with respect to compliance with CALEA, the statute that addresses cooperation with law enforcement. Packet-switched technology poses unique technical issues, but manufacturers and providers of VoIP are moving ahead to implement compliance capabilities into their systems. Moreover, CALEA has a different definition of telecommunications than the Communications Act, so there is no need to define VoIP as telecommunications for Communications Act purposes in order to mandate that VoIP manufacturers and service providers cooperate with law enforcement.

- **Universal Service.** As for universal service, of course, VoIP providers directly or indirectly already contribute to USF. The fact that more and more calls, including wireless and business calls made on modal access as well as some VoIP calls, don't contribute or contribute unevenly to USF should not be an excuse for regulation of all these modes. Instead, what is needed is reform of funding for explicit USF. We believe that a numbers-based contribution mechanism would better ensure the continued sustainability of USF than any attempt simply to include VoIP or other information services in the current revenue-based mechanism. If one of the goals of universal service is to provide affordable voice communications to rural America, then no technology offers more promise for providing more affordable communications, not only to rural America, but to all of America.

- **Inter-carrier compensation.** We urge the FCC to move away from a hodgepodge of implicit subsidies and towards a rational series of voluntary inter-carrier business arrangements with regulation required only when there is effective monopoly ownership of a bottleneck. "Bill and keep" may well turn out to be an effective arrangement as it has been in much of the IP world.

- **Phone-to-Phone VoIP Regulation.** One suggestion that has been made is that phone-to-phone Voice over IP be regulated while "other" VoIP is not. This would be a mistake even if it were possible and it is, in fact, impossible to define today what is a phone. It is phone-to-phone traffic which has funded and continues to fund the buildout of a worldwide network of interfaces between the PSTN and the Internet around the world. These interfaces are necessary so that VoIP phone and voice PBXes can connect with the TDM world and vice versa. It is the existence of these networks of traffic exchange points which are making possible the deployment of innovative new VoIP services because the users of these services have full connectivity to the TDM world – not just to other VoIP users.

- **State Role.** We also don't deny that there is a legitimate role for state governments, but that role has to be defined in a way that is consistent with the interstate nature of the Internet and the practical problems that would be caused by varying state regulation.

- **FCC Role.** We believe that the FCC has the legal authority to continue to keep its hands off the Internet and IP networks even when they are used for voice applications. Voice over IP should be classified as an information service and regulated only to the extent necessary pursuant to the Commission's Title I or ancillary jurisdiction.

VoIP providers shouldn't be regulated like phone companies with large market power. The historic reason for telephony regulation was the existence of monopoly providers and an infrastructure which made it nearly impossible to challenge such monopolies even in the rare case where it was legal to do so. In contrast, a provider of a VoIP service has no need to own or build the infrastructure on which the service is delivered and, since there are no historic or even nascent VoIP monopolies, there is simply no basis for regulation of any such provider that does not have significant market power.

Von Coalition Members			
AT&T	**AT&T** provides a residential broadband Internet voice service called CallVantage. AT&T also provides enterprise IP Services as well as phone based VoIP services. http://www.att.com/	MCI	**MCI**, with headquarters in Ashburn, VA, is a leading provider of global communications services and operates an expansive IP backbone. http://www.mci.com
BMX	**BMX**, based in New York, is a competitive provider of call center applications using Internet-based telecommunications services. http://www.bmx-inc.com/	Microsoft	**Microsoft**, based in Redmond Washington, is a leading VoIP innovator for software, embedded systems, devices and enterprise applications. http://www.microsoft.com/
Callipso	**Callipso** provides enhanced IP services to telecommunications carriers, enterprises and the prepaid card market. It is based in Santa Ana, CA. http://www.callipso.com/	POINTONE	**PointOne** is a VoIP network provider and offers IP communications services to the provider community. It is based in Austin, TX. http://www.pointone.com/
CallSmart	**CallSmart** is a North Dakota company that is taking advantage of the latest Voice over Internet (VoIP) technology to provide an alternative to traditional long distance voice service. http://getcallsmart.com/	pulver.com	**Pulver.com**, based in Melville, New York, is an early VoIP innovator and offers the Free World Dialup service which provides consumers Free Telephony over Broadband. http://www.pulver.com/
CONVEDIA	**Convedia** Corporation is based in Vancouver, Canada. It is a supplier of next-generation IP media services. http://www.convedia.com/	Skype	**Skype** is a Global P2P Telephony Company that is offering consumers free, superior-quality calling worldwide. http://www.skype.com/
COVAD	**Covad**, based in San Jose California, is a leading national broadband service provider of high-speed Internet and network access utilizing Digital Subscriber Line (DSL) technology. http://www.covad.com/	TELEGLOBE	**TeleGlobe** is a leading provider of international voice, wireless roaming and data/IP services. http://www.teleglobe.com
iBasis	**iBasis**, based in Burlington, MA provides wholesale international telecommunications services, and has carried more than 6 billion minutes of international voice traffic. http://www.ibasis.com/	Texas Instruments	**Texas Instruments** of Dallas, TX is a leader in digital signal processing and analog technologies to meet signal processing requirements and support the development of IP telephony. http://www.ti.com/
IceNet	**IceNet** of Dallas, TX, provides VoIP infrastructure to VoIP Application Service Providers and Broadband Providers for enhanced local, long distance, toll-free and DID services. http://www.myicenet.com/	VocalData	**VocalData**, based in Richardson, Texas, provides hosted IP telephony applications that enable service providers to reliably and cost-effectively deliver voice-over-IP solutions. http://www.vocaldata.com/
intel	**Intel** is the world's largest chip maker. Based in Santa Clara, CA, it also manufactures computer, networking and communications products. http://www.intel.com/	voiceglo	**VoiceGlo**, headquartered in Fort Lauderdale, Fla, is a global, full-service Voice over Internet Protocol (VoIP) communications company. http://www.voiceglo.com/
intrado	**Intrado**, based in Longmont, CO., provides emergency service solutions to Enable VoIP 9-1-1 calls, to the public safety and telecommunications industries. http://www.intrado.com/		